# The Crash Outcome Data Evaluation System (CODES) And Applications to Improve Traffic Safety Decision-Making

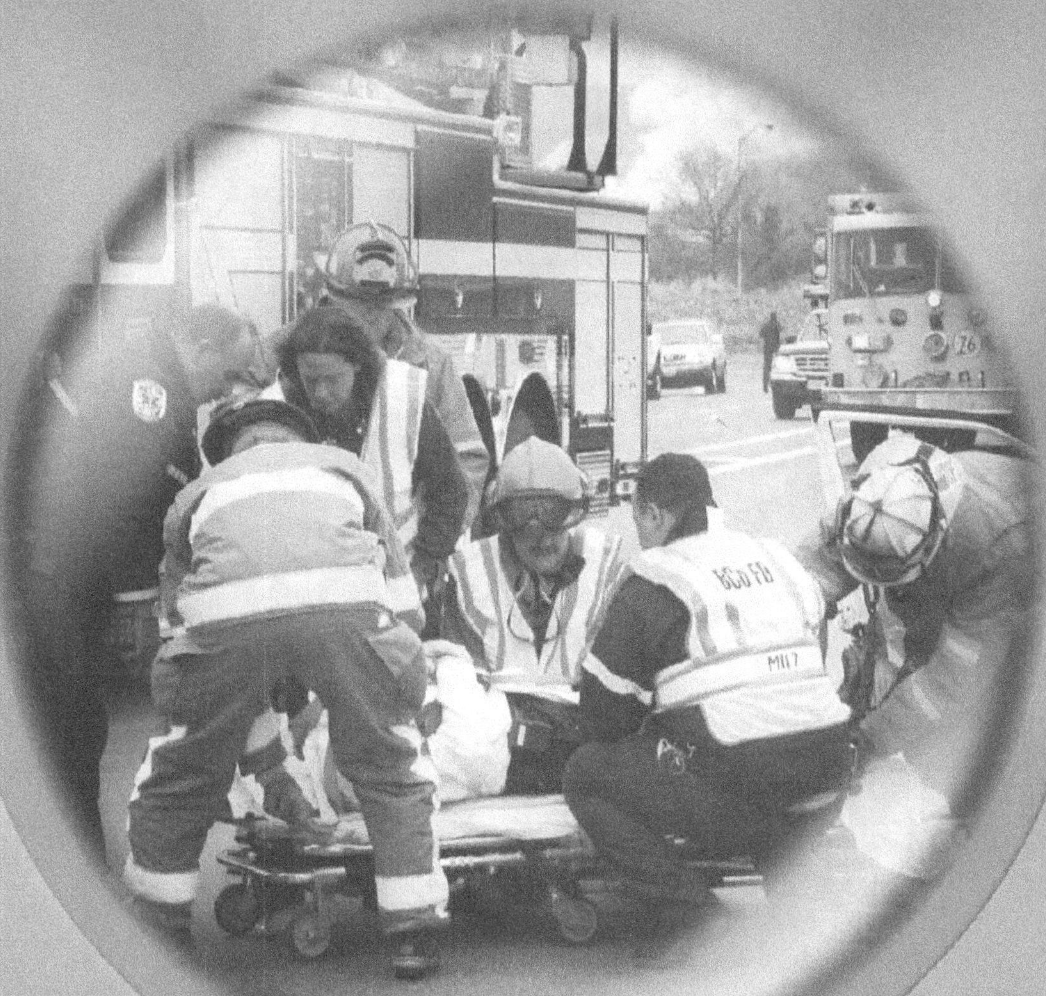

**NHTSA**
www.nhtsa.gov

Link**a**ge

The Crash Outcome
Data Evaluation System

1
1
1

# CONTRIBUTING AUTHORS

Alabama: Glenn Cummings, Center for the Study of Rural Vehicular Trauma
Connecticut: Justin Peng and Marian Storch, Connecticut Department of Public Health
Delaware: Steven Blessing and Laurie Lin, Delaware Office of EMS
Georgia: Denise Yeager, Georgia Dept. of Human Resources./Div. of Public Health/Injury
          Prevention
Illinois: Mehdi Nassirpour, Illinois DOT Division of Traffic Safety
Indiana: Jose Eduardo Thomaz, Center for Road Safety, Purdue University
Iowa: Suning Cao, Iowa Dept. of Public health/Center for Vital Records and Health Statistics;
          Scott Falb, Office of Driver Services/Iowa Dept. of Transportation
Kentucky: Huifang (Jenny) Qin and Michael Singleton, Kentucky Injury Prevention and
Research Center
Maine: Karl Finison, Maine Health Information Center
Maryland: Timothy Kerns, University of Maryland/National Study Center for Trauma and EMS
Massachusetts: Heather Rothenberg, University of Massachusetts
 Minnesota: Tina Folch and Scott Hedger, Minnesota Dept. of Public Safety/Office of Traffic
          Safety; Anna Gaichas and Mark Kinde, Minnesota Dept. of Health
Missouri: Mark Van Tuinen, Missouri Department of Health and Senior Services
Nebraska: Ming Qu, Nebraska Department of Health and Human Services
New York: Motao Zhu and Susan Hardman, New York State Department of Health
Ohio: Kristen Conner and Dr. Gary Smith, Children's Research Institute/Ohio State University
Rhode Island: Ted Donnelly, Rhode Island Department of Health
South Carolina: Mary Tyrell and Tracy Joyce, South Carolina State Budget and Control Board
Utah: Larry Cook, Intermountain Injury Control Research Center, University of Utah
Virginia: Michael Lundberg and Kathleen Bernard, Virginia Health Information


From NHTSA's National Center for Statistics and Analysis:
Augustus "Chip" Chidester, Director, Office of Data Acquisitions
Barbara Rhea, Chief, State Data Reporting Systems Division
John Kindelberger, Team Leader, State Data and Quality Assurance
Xuemei Pan, Advanced Systems Technology and Management, Inc., for NHTSA/NCSA
Morrie O'Neil, Advance Systems Technology and Management, Inc.

Governors Highway Safety Association for NHTSA/NCSA:
Sandra Johnson
Michael McGlincy, Ph.D., Strategic Matching Inc.

TABLE OF CONTENTS

# EXECUTIVE SUMMARY

The Crash Outcome Data Evaluation System (CODES) is a program facilitated by the National Highway Traffic Safety Administration as a component of its State Data Program. CODES uniquely uses probabilistic methodology to link crash records to injury outcome records collected at the scene and en route by emergency medical services, by hospital personnel after arrival at the emergency department or admission as an inpatient and/or, at the time of death, on the death certificate.

CODES is designed to foster and cultivate crash-outcome data linkage for highway safety applications at the State level, supporting State Highway Safety Offices, State Public Health and Injury Prevention Departments, State Emergency Medical Services Agencies, State transportation departments, and other such agencies; and to facilitate participation in NHTSA-coordinated multistate studies using linked data at the Federal level. This document is intended to inform traffic safety professionals, from those in CODES programs to those in the agencies they support, as well as all others interested in traffic safety, on best-practice applications available through linked CODES data.

To support CODES program objectives, NHTSA sponsors cooperative agreements that provide software access, technical assistance, and program assistance to CODES State programs to link information about State-reported crashes and their consequences, and to analyze and disseminate the information. Analyses of linked data help inform State traffic safety professionals and coalitions to determine and implement data-driven traffic safety priorities

CODES evolved from a need to quantify and report on the benefits of safety equipment and legislation in terms of mortality, morbidity, injury severity, and health care costs, and has built proactive partnerships between the traffic safety and public health agencies, which own the State data, and NHTSA, which provides access to the software and training resources that make the linkage feasible. NHTSA maintains a CODES facilitating infrastructure to provide technical support and assistance to sites while also encouraging each site to build State-level collaborations and perform relevant analyses within their States.

Topics of interest addressed by CODES have included pre-hospital, emergency department, inpatient, rehabilitation, and other health care charges by payer source (private, workers' compensation, Medicare, Medicaid, etc.), and associations with the consequences of motor vehicle crashes; crash injury patterns by type and severity, and hospital charges, by such variables as safety equipment use, vehicle type, geographical location, and others. In recent years, as outlined in abstracts in this paper, such information has been used to:

- *Identify Traffic Safety Problems:*

    CODES data has been used to identify traffic safety issues in numerous ways, which include examining whether the increased crash rates for teen drivers is accompanied by an increased injury to their passengers (New York and Minnesota); determining hospital charges and length of hospital stay for motorcycle-related injuries (Georgia); identifying the effect of seat belt usage in preventing injuries and fatalities (Kentucky, New York, Ohio, and Utah); studying injury patterns among children riding with unbuckled adults

1

compared to buckled adults (Utah); researching the types and frequency of injuries to children in passenger motor vehicles (Connecticut and Missouri); and analyzing the characteristics and outcomes of crashes involving teen drivers (Delaware and Minnesota).

- ***Support Traffic Safety Decision-Makers:***

  CODES data has been used to inform and educate traffic safety decision-makers as the State and local level. Examples of CODES activities being used to support decision-makers include providing State legislators with the CODES report on the effectiveness of seat belt use in preventing injuries and fatalities (Kentucky); delivering data and expertise to the State Highway Administration to assist in the development of a long-term, statewide strategic plan to guide the future direction of traffic records and highway safety (Maryland); presenting CODES-related fact sheets and reports to the State Traffic Records Coordinating Committee (TRCC) (Massachusetts, Virginia); using CODES data to present a report to the Governor's Highway Safety Office and State legislators on the effect that enacting a standard enforcement seat belt law would have on hospital charges, direct medical costs, and the impact to the State's Medicaid system (Ohio); and developing media products on the medical and financial consequences of the under-the-influence drivers involved in crashes (South Carolina).

- ***Support Traffic Safety Legislation:***

  CODES research has been used at the State level to inform legislators about traffic safety issues in their State and traffic safety legislation. These activities included providing legislators with information in support of upgrading existing graduated driver's license (GDL) laws (Delaware and Minnesota); using CODES data to expand a mandatory seat belt law to include back-seat passengers (Indiana); creating a fact sheet to help support legislation for motorcycle helmet use (Iowa); and providing data to a children's safety advocacy group to help convince legislators to change child passenger safety (CPS) laws (Connecticut).

- ***Educate the Public:***

  As a means to informing the public about traffic safety issues, CODES data has been used to give a presentation to a State TRCC about the length of hospital stay, injury body region, and nature of injury for older vehicle occupants (Massachusetts); post a fact sheet comparing the crash rate severity of State drivers against non-State drivers on the State's Health and Human Services Department Web site (Nebraska); publish a paper on backseat seat belt use in the *Journal of Safety Research* (New York); and launch a Web site that contains five years of CODES data from which the user can select standard reports or create online queries based on selected criteria (Virginia).

NHTSA is also using CODES multi-State data in a variety of studies as CODES States enable submission of standardized, non-identifiable data for research purposes. Through streamlining of programs and continuing methodological innovations, CODES is expected to continue leadership on traffic safety research in the 21st century.

# I. THE CRASH OUTCOME DATA EVALUATION SYSTEM

## Introduction and Background

Evolving from a need to quantify and report on the benefits of safety equipment and legislation in terms of mortality, morbidity, injury severity, and health care costs at State and national levels, the Crash Outcome Data Evaluation System has built proactive partnerships between the traffic safety and public health agencies, which own the State data, and NHTSA, which provides access to the software and training resources that make the linkage feasible. NHTSA maintains a CODES facilitating infrastructure to provide technical support and assistance to sites while also encouraging each site to build State-level collaborations and perform relevant analyses within their States.

The intent of CODES data linkages is to ensure that traffic safety coalitions have access to crash outcome analyses to help determine and implement data-driven traffic safety priorities. As a result of these linkages, the availability of population-based, comprehensive, and representative crash outcome data is maintained to assist existing traffic safety coalitions in the selection and implementation of data-driven traffic safety priorities. A properly implemented State CODES program supports State Highway Safety Offices, State Public Health and Injury Prevention departments, State EMS Agencies, State transportation departments, and other such agencies to target their resources and evaluate the potential impact on preventing mortality and morbidity, reducing injury severity, and lowering health care costs.

## Implementation and Data Enhancement

Through NHTSA-sponsored cooperative agreements, CODES provides access to linkage software and technical assistance in the form of software support, linkage training, and analysis training to State CODES programs. Technical assistance is disseminated through means such as software documentation, Web seminars, individual assistance, and periodic technical assistance meetings. Once trained, State CODES programs add to the available knowledge about State-reported crashes and their consequences by linking crash data to data systems such as EMS records and statewide hospital discharge records.

The CODES program links crash records to injury outcome records collected at the scene and en route by emergency medical services, by hospital personnel after arrival at the emergency department or admission as an inpatient and/or, at the time of death, on the death certificate. Although crash data indicate the occurrence of injury, they include only limited information about type and severity and no information about health care costs or International Classification of Diseases (ICD) coding. Similarly, hospital injury datasets with ICD codes and billing charges do not include information about the characteristics of the crash or vehicles involved. CODES linkage integrates the two subject matters, and as a result, CODES provides statewide, real-world crash outcome data that can provide population-based information on crash outcomes in terms of deaths, specific injury type/region/severity, and costs.

As linkage expands into other data systems, as it has in some States, it can enhance other participating data systems in a variety of ways. For example, EMS and hospitals obtain information about the time of onset to evaluate the responsiveness of the trauma system;

roadway inventories expand to include injury outcome information by location point; driver licensing information is augmented with the medical and financial consequences caused by drivers who are impaired or repeat offenders; and vehicle characteristics can be related to specific types of injuries and their costs. In addition, data quality can improve as previously undiscovered problems are identified and corrected during linkage.

**Federal Use of CODES Data**

In the original NHTSA Report to Congress (1996) on the Benefits of Safety Belts and Motorcycle Helmets, data on crashes from seven States was compiled as part of the first CODES program. This report employed probabilistic linkage techniques to combine data gathered from police crash reports, emergency medical services, hospital emergency departments, and hospital discharge files to more fully describe motorcycle crash events and their outcomes. Among other findings, this report revealed an increase in hospital charges for motorcycle operators who were unhelmeted at the time of their crashes, and motorcycle helmet effectiveness of 67 percent in the reduction of brain injuries.

Though not catalogued in this report, the Federal-State collaboration that successfully implemented CODES at the State level has led to the planning of new CODES applications at the Federal level. States that have successfully linked at least two years of crash and injury outcome data may receive the benefits of the NHTSA CODES facilitation as part of the CODES Data Network. These projects also share their expertise and, under certain privacy considerations, can contribute specific standardized model variables for multi-State studies. NHTSA is currently working with CODES Data Network States to expand this capability, in order to provide support to NHTSA program needs with multi-State crash data analyzed by crash conditions, safety equipment, and other crash variables in terms of injury types, level of care, discharge status, payer, charges, and other outcome variables not available in unlinked crash data.

**Document Purpose and Structure**

This document is designed to well-inform traffic safety professionals, from those in CODES programs to those in the agencies they support, as well as all others interested in traffic safety, on best-practice applications available through linked CODES data. To distribute information on many applications in a compact way, this paper presents abstracts of applications presented by CODES members at CODES annual meetings. Abstracts are presented as prepared for the years 2006 and 2007. Each summarizes the population and traffic safety issue targeted, the format and methodology used, and impacts, actions or follow-ups on the targeted subject matter.

Since the State-specific applications presented are limited to those reported at the CODES meetings in 2006 and 2007, they do not represent a complete inventory of all of the applications developed by the States during this period. Thus, for example, not all of the CODES States actively involved in developing and supporting the State TRCC are listed in Table 2.

For convenience, abstracts are summarized in four categories, with summary tables allowing reference and page location by category and subject matter. After the summaries, full abstracts are presented in State alphabetical order. Following the abstracts, two appendices offer responses to frequently asked questions regarding the CODES program and a more technical description of the statistical methodologies used to achieve representative linked data statewide.

**CODES State-Level Applications Summarized**

This section summarizes presented State CODES applications in four broad objectives: (1) traffic safety problem identification, (2) traffic safety decision-making support, (3) safety legislation development and support, and (4) public education. For each grouping, a table summarizes and references State applications in the category.

Objective 1: Identify Traffic Safety Problems
CODES data is population-based, so the large volume of data generated as the result of annual linkage in most CODES States can help identify safety issues including infrequent but potentially significant crash outcomes. Table 1 provides a reference to traffic-safety problem-identification applications as documented in their abstracts in Part 2 of this report.

Objective 2: Support Traffic Safety Decision-Makers
With limited resources, decision-makers need to identify and justify priorities assigned to improving traffic safety in relation to other public health issues. CODES data can provide statewide information to support safety efforts initiated by elected officials, and CODES data reported at the county or local level can be used in planning priorities of the coordinating agencies such as the Traffic Records Coordinating Committees (TRCCs), funding/planning agencies such as the State Highway Safety Offices and departments of public health/injury control, or data users such as the members of the CODES advisory committee. Many CODES projects also have played key roles in helping to define the traffic safety agenda of their State TRCC committees. Table 2 summarizes examples of these traffic-safety decision-making support applications as documented in their abstracts in Part 2 of this report.

Objective 3: Support Traffic Safety Legislation
Traffic safety efforts must be targeted where they will have the most impact on improving crash outcome. When the increase in deaths, injury, severity, and health care costs becomes unacceptable, legislation may be necessary to change public behavior. In these instances, the State-specific medical and financial information generated by CODES can demonstrate the expected savings, in terms of decreased health care costs and injury severity, to taxpayers with the adoption of specific traffic safety legislation. Table 3 summarizes examples of legislation-support applications as documented in their abstracts in Part 2 of this report.

Objective 4: Educate the Public
If an educated public can help improve traffic safety, then traffic safety information must be readily available in a format that meets the public's needs. In support of public education, State CODES programs often use 1– or 2-page fact sheets to disseminate traffic safety information to the public.. In addition, the Internet has increased accessibility to linked crash outcome results while complying with State privacy legislation and regulations. Some States provide online interactive reports or detailed data tables. Table 4 summarizes examples of public education applications as documented in their abstracts in Part 2 of this report, and Table 5 lists the Web sites for the CODES States and NHTSA as of March 2009.

**Table 1: State CODES Applications Supporting Traffic Safety Problem Identification**

| State and Application by Type of Subject Matter | | Page |
|---|---|---|
| **1. Aggressive Drivers** | | |
| Delaware | **Aggressive Driving Study**<br>Descriptive statistics and rate ratios (RRs) were used to compare the crash outcomes between 16- and 17-year-old drivers with and without passengers. | 18 |
| **2. Older Drivers** | | |
| Rhode Island | **Elder Occupants in Motor Vehicle Crashes: Forecasting Health Burden**<br>Statistically, Rhode Island's older occupants are more likely to be hospitalized or fatally injured after motor vehicle crashes. CODES data was used to examine the differences and the age distribution of the population to help in developing planning policies on future highway safety interventions. | 50 |
| **3. Teen Drivers** | | |
| Minnesota | **The Epidemiology of Motor Vehicle Crashes Involving 16- and 17-Year-Old Drivers in Minnesota and Associated Hospital Charges**<br>CODES data were used to determine the per mile rate of teenage driver motor vehicle crash involvement and the characteristics of these crashes in Minnesota. Measure medical care charges and the severity of injuries associated with motor vehicle crashes involving teenage drivers. | 38 |
| New York | **Using Multiply-Imputed CODES Data to Identify Risk Factors and Reveal Societal Costs in Teen Driving**<br>Police crash reports, emergency department and hospital discharge data were examined to determine the risk factors and societal costs for drivers age 16 to 20. Teen drivers and 25- to 49-year-old drivers were compared for traffic crash and injury rates, emergency department visit rates, hospitalization rates, and crash contributing factors. | 46 |
| **4. Non-Resident Drivers** | | |
| Nebraska | **Comparing Crashes That Occurred in Nebraska Involving Nebraska and Non-Nebraska Drivers**<br>Comparing the crashes occurring in Nebraska from 1999 to 2003 by the driver's State of residence, CODES data explored the patterns and the contributing risks factors of the crashes. Crashes involving non-Nebraska drivers tended to be more severe, resulting in more deaths and serious injuries than crashes involving Nebraska drivers. | 44 |
| **5. Roadway** | | |
| Nebraska | **Why Is It So Risky to Drive on the Roadways Where the Posted Speed Limit Is 50 Miles Per Hour?**<br>The crashes that occurred on roadways with posted speed limits of 50 mph resulted in severe crash outcomes with the highest injury rate and a higher death rate than those crashes that took place on other roadways. This CODES study examines the causes and consequences of crashes occurring on Nebraska roads with posted speed limits of 50 mph. | 43 |
| **6. Traumatic Brain Injury** | | |
| Iowa | **Motor Vehicle Crash (MVC) Related to Traumatic Brain Injuries in Iowa**<br>To demonstrate the growing crash and injury rates among motorcycle riders in Iowa from 2001–2005, CODES data was used to examine the five years' crash rates per 1,000 motorcyclist licensed drivers—including fatality and injury rates— and were calculated to demonstrate low rates of helmet use in Iowa. | 23 |
| **7. Data Quality Issues** | | |
| Maine | **Adjusting for Seat Belt Reporting: The Problem of Differential Misclassification**<br>CODES reviewed the problem of differential misclassification of seat belt use in police crash reports and explored a method of correcting for missing data and differential misclassification using CODES linked data. | 30 |

**Table 2: State CODES Applications Supporting Traffic Safety Decision-Makers**

| State and Application by Type of User | | Page |
|---|---|---|
| **1. Elected Officials: Governor and Legislators** | | |
| Missouri | *Safety Device Use for Children Age 4 to 8*<br>CODES data was used to measure the effect of safety device use on emergency department (ED) and inpatient charges, ejection, hospitalization, traumatic brain injury (TBI), and death for children age 4 to 8. | 40 |
| South Carolina | *Providing Information to Support Decision Making*<br>CODES data was used to develop various media products distributed to members of the South Carolina legislature and traffic safety decision makers. | 51 |
| **2. Coordinating Agencies: Traffic Records Coordinating Committee and Strategic Planning** | | |
| Connecticut | *The Importance of CODES Involvement in the Connecticut TRCC*<br>Content about the CODES project was submitted as part of the Connecticut TRCC's 5-Year Strategic Plan. The Strategic Plan identified CODES as one of the top priorities for State traffic records improvements. The needs of CODES, as highlighted in the Strategic Plan, will help advocate for the program both within the Connecticut Department of Public Health (DPH) and statewide. | 13 |
| Iowa | *Iowa Motor Vehicle Crash Injury Facts Among Emergency Department Visits: A Profile of Unbelted Occupants of Passenger Cars and Vans/SUVs (2003)*<br>CODES data analysis was presented at a TRCC meeting and to the Prevention of Disability Council Meeting. The data was also used for occupant restraint task force group within the TRCC. | 24 |
| | *Motorcycle Quick Facts in Iowa – 2001 to 2005*<br>Helmets were evaluated and a fact sheet was distributed at the Iowa Motorcycle Safety Forum 2007. The data results have been included in Iowa's Safety Management Systems within Traffic Safety and are being used to support Iowa goals. | 25 |
| Maryland | *Using Data for Strategic Planning and TRCC*<br>The CODES project has contributed data and expertise to the State Highway Administration to assist in the development of a statewide strategic plan that will guide the direction of traffic records and highway safety in the State for the next several years. | 31 |
| Massachusetts | *Crash Outcomes for Older Occupants of Motor Vehicles in Crashes*<br>CODES Data results were presented in several formats including fact sheets, reports, and presentations to the Massachusetts CODES advisory board and Massachusetts Traffic Records Coordinating Committee. | 35 |
| **3. Funding Agencies: Highway Traffic Safety Offices, Strategic Planning** | | |
| Massachusetts | *Evaluation of Frequency and Injury Outcomes of Lane Departure Crashes*<br>Lane departure data analysis has been an ongoing project for MassHighway and the dialogue between the MassHighway safety analysts and UMassSAFE have included a discussion of how the CODES data can be used and the benefits of using CODES data in addition to traditional crash data. | 33 |
| Virginia | *408 Funding Award for Virginia's Traffic Safety Information System*<br>CODES staff developed a study for the Governor of Virginia, members of the Virginia General Assembly, the Traffic Records Coordinating Committee, the Highway Safety Office, and the Injury Control Office. | 56 |
| **4. Users of Traffic Safety Data: CODES Advisory Committee, Injury Control** | | |
| Illinois | *Inpatient Charges and Utilization Patterns of Crash Victims in Illinois*<br>Preliminary CODES findings have been presented to the Illinois CODES Advisory Group and made available to the Illinois Traffic Records Coordinating Committee. | 21 |
| Ohio | *Boost Advisory Board Interest in CODES*<br>The CODES team used input from the advisory board to focus on topics of interest to the traffic safety community. | 47 |

**Table 3: State CODES Applications Supporting Traffic Safety Legislation**

| State and Application by Type of Issue | | Page |
|---|---|---|
| **1. Convert Secondary to Primary Safety Belt Law** | | |
| Minnesota | *Estimating Minnesota Hospital Charge Savings With the Adoption of a Standard Enforcement Seat Belt Law*<br>A CODES report on seat belts was used to gather support in the Minnesota Senate for passage of a primary seat belt bill. | 37 |
| Ohio | *The Impact of a Standard Enforcement Safety Belt Law on Fatalities and Hospital Charges in Ohio: An Analysis Using 2003 Ohio CODES Data*<br>The Ohio Governor's Highway Safety Office at ODPS assisted in publicizing and sharing CODES results with the Governor and other legislators. | 48 |
| Utah | *Primary Safety Belt Enforcement Efforts*<br>A fact sheet titled "The Cost of Being Unbuckled," was given to every State legislator. | 54 |
| **2. Expand Primary Belt Legislation to Include Backseat Passengers** | | |
| New York | *The Importance of Promoting Backseat Safety Belt Use*<br>A presentation was made during the New York State Highway Safety Conference and a paper on backseat seat belt use was published in Journal of Safety Research. | 45 |
| **3. Expand Primary Belt Legislation to Include Pickup Trucks** | | |
| Georgia | *Effects of Pickup Trucks Seat Belt Exemption*<br>Georgia CODES provided data to support the House bill to eliminate the current exemption in Georgia code pertaining to the use of seat belts in pickup trucks. | 20 |
| Indiana | *Effect of Seat Belt Usage on Hospital Charges by Vehicle Type*<br>CODES research was used to support House Bill 1237, which now requires all occupants, including backseat passengers in all vehicles to buckle up. | 22 |
| **4. Improve Child Passenger Safety** | | |
| Connecticut | *Usefulness of CODES Data for CODES Advisory Board Members and Using CODES Data to Improve Connecticut's Child Passenger Safety Law*<br>Staff met with the Director of Connecticut Safe Kids, a member of the advisory board, to identify analyses and presentation formats that will be most effective in supporting improvements to Connecticut's child passenger safety laws. | 15 |
| Kentucky | *Economic Costs of Low Safety Belt Usage in Kentucky*<br>The Kentucky legislature passed a primary seat belt enforcement bill in the General Assembly. The Kentucky CODES report was one of the efforts that preceded its passage. | 26 |
| **5. Strengthen Graduated Driver Licensing** | | |
| Delaware | *Support for Passenger Limitations on 16- and 17-Year-Old Drivers*<br>A CODES fact sheet was forwarded to legislative liaisons to support upgrading Delaware's Graduated Driver's License (GDL) law. | 17 |
| Utah | *Using CODES Data to Strengthen Utah's GDL Law*<br>CODES findings were published in several newspapers and aired on local television news programs. As a result, local traffic safety advocates asked Utah CODES to prepare fact sheets about GDL, which were given to legislators and presented in committee meetings. | 53 |
| **6. Reduce DUI** | | |
| Rhode Island | *Rhode Island CODES Provides Expert Participation*<br>Rhode Island's CODES linked data results and other highway safety and public health information has been used in support of stricter laws against driving under the influence. | 49 |
| South Carolina | *The Continuing Saga of DUI Legislation*<br>Media products are distributed to legislators and traffic safety decision makers to illustrate the medical and financial consequences of under-the-influence drivers involved in crashes. The data are analyzed at a local level to show local consequences. | 52 |
| **7. Support Helmet Legislation** | | |
| Georgia | *Motorcycles in Georgia*<br>Fact sheets on motorcycle crash injuries in Georgia were developed for different audiences during unsuccessful efforts to repeal Georgia's motorcycle helmet law. | 19 |

**Table 4: State CODES Applications Supporting Public Educations**

| State and Application by Subject or Format | Page |
|---|---|
| **FACT SHEETS** | |
| **1. Motorcycles** | |
| Georgia<br>***Motorcycles in Georgia***<br>Two fact sheets were developed by Georgia CODES staff to inform policy makers and the general public about motorcycle crashes and motorcycle hospitalizations. | 19 |
| Iowa<br>***Motorcycle Quick Facts in Iowa 2001–2005***<br>A fact sheet was created by Iowa CODES staff and distributed at the Iowa Motorcycle Safety Forum. | 25 |
| **2. Seat belts** | |
| Georgia<br>***Safety Belts and Pickup Trucks***<br>A fact sheet was developed by Georgia CODES staff to inform policy makers and the general public about issues relating to seat belts and pickup trucks. | 20 |
| Minnesota<br>***Estimating Minnesota Hospital Charge Savings With the Adoption of a Standard Enforcement Seat Belt Law***<br>A CODES report and fact sheet were distributed through statewide e-mail distribution lists to law enforcement, public health educators, and other traffic safety advocates. The report was highlighted on the Office of Traffic Safety and Department of Health's Web sites. CODES data will continue to be used in various publications and presentations that discuss the economic impact of crashes. | 37 |
| **3. Teen Drivers** | |
| Delaware<br>***Passenger Limitations for 16- and 17-Year-Old Drivers***<br>CODES staff created a fact sheet that highlighted CODES data and supports passenger limitation for 16- and 17-year-old drivers. | 17 |
| Kentucky<br>***Injuries to Booster-Age Children in Kentucky***<br>Kentucky CODES created a fact sheet that focused on the outcomes (injury type, hospital charges, hospital length of stay, etc.) for restrained and unrestrained children age 4 to 8, compared with other age groups. | 27 |
| **4. Non-Nebraska Drivers** | |
| Nebraska<br>***Non-Nebraska Drivers Involved in Motor Vehicle Crashes Occurring in Nebraska***<br>A fact sheet was created by Nebraska CODES staff and posted on the Nebraska Health and Human Services System's Web site. | 44 |
| **DATA ACCESS VIA THE INTERNET** | |
| Missouri<br>***Update of Web Application***<br>The Missouri CODES Web application allows users to develop tables of CODES data that relate to seat belt/helmet use, alcohol involvement, driver gender and age, crash type, speed zone, and vehicle type to level of medical care/death, ejection, and TBI. Five years of CODES data are now on the Web site | 41 |
| Virginia<br>***Virginia CODES Web site***<br>Virginia's CODES Web site, www.vacodes.org, was launched in May 2006 and currently contains five years of data from which the user can select standard reports or create online queries based on selected criteria. | 55 |

**Table 5: Web Sites of CODES Programs in the NHTSA CODES Network**

| State | Web Address (current March 2009) |
|---|---|
| Delaware | www.dhss.delaware.gov/dhss/dph/EMS/codeshome.html |
| Indiana | http://cats.ecn.purdue.edu/CODES.aspx |
| Iowa | http://www.idph.state.ia.us/apl/codes.asp |
| Kentucky | www.kiprc.uky.edu/projects/CODES |
| Maine | www.mhic.org/CODES |
| Maryland | http://medschool.umaryland.edu/NSCforTrauma//codes.asp |
| Massachusetts | www.ecs.umass.edu/umasssafe/codes.htm |
| Minnesota | www.health.state.mn.us/injury/topic/topic.cfm?gcTopic=9 |
| Missouri | www.dhss.mo.gov/MICA/index.html **(Select motor vehicle crashes)** |
| Nebraska | http://www.hhss.ne.gov/codes/ |
| Ohio | http://sharedoc.nchri.org/CIRP/Pages/CODES.aspx |
| Rhode Island | www.health.ri.gov/chic/statistics/codes.php |
| South Carolina | www.ors2.state.sc.us **(select SC CODES project)** |
| Utah | http://www.utcodes.org |
| Virginia | www.vacodes.org |

## II. APPLICATIONS TO IMPROVE TRAFFIC-SAFETY DECISION-MAKING

To disseminate exemplary data uses and promising practices throughout the CODES Data Network and beyond, abstracts of presentations given at CODES annual meetings are presented on the following pages. Abstracts are in alphabetical order by State, and each one describes an application that was presented at a national CODES Technical Assistance meeting held in either 2006 or 2007. At these meetings, State CODES programs are encouraged to share ideas and progress in the program realm as well as the technical areas. The abstracts are offered as an exemplary yet partial representation of programmatic and technical CODES endeavors and achievements undertaken in recent years.

# Alabama CODES
## Center for the Study of Rural Vehicular Trauma
State Application Presented at CODES Annual Meeting, 2006
## Title: *Does Increased EMS Pre-Hospital Time Affect Patient Mortality In Rural Motor Vehicle Crashes? A Statewide Analysis.*

Abstract: The purpose of this study is to link and analyze Alabama's statewide pre-hospital data and identify whether EMS pre-hospital time affects patient mortality from vehicular trauma.

| | |
|---|---|
| Contact Person and number | Glenn Cummings, 251-471-7501<br>Director, Center for the Study of Rural Vehicular Trauma |
| Population Targeted | Rural vehicular trauma patients |
| Purpose | Reduce mortality rates due to rural vehicular trauma |
| Type of Decision Maker Targeted | Alabama Department of Public Health, Division of Emergency Medical Services |
| Description of linked data and analyses | Linkage of data from police motor vehicle crash and Emergency Medical Services Patient Care Reports |
| Analytical Results | Pre-hospital data was analyzed to determine EMS response times, scene times, and transport times in rural and urban settings. During a two-year period (January 2001 – December 2002) 34,341 people (75%) and 11,422 people (25%) were injured in rural and urban settings, respectively. The data also showed that 611 (1.78%) fatalities occurred in rural settings, and 103 (.90%) fatalities occurred in urban settings ($p<0.001$). The overall mean EMS pre-hospital time when fatalities occurred was 42.0 minutes in the rural setting versus 24.8 minutes in the urban setting ($p<0.001$). |

| Pre-Hospital Times for ll Fatalities (min) | Rural | Urban | p-value |
|---|---|---|---|
| Mean EMS Response Time | 10.67 | 6.50 | $< 0.0001$ |
| Mean EMS Scene Time (excludes DOS and extrication) | 18.87 | 10.83 | $< 0.0001$ |
| Mean EMS Transport Time (excludes DOS) | 12.45 | 7.43 | $< 0.0001$ |
| Overall Mean EMS Pre-Hospital Time | 42.0 | 24.8 | $< 0.0001$ |

| | |
|---|---|
| Dissemination Formats | 1) Presentation at the American Association for Surgery of Trauma.<br>2) Expected publication in the Journal of Trauma<br>3) Presentation to the Alabama Department of Public Health, State Emergency Medical Control Committee<br>4) Presentation at the Alabama Chapter of the American College of Surgeons. |
| Impact, Follow-Up, or Later Development on the Targeted Issue | Alabama's Department of Public Health, Division of Emergency Medical Services, has suggested further investigation into the cause of increase in response times, scene times, and transport times in rural settings. |

| | |
|---|---|
| | <div align="center">**Connecticut CODES**<br>**Connecticut Department of Public Health**<br>State Application Presented at CODES Annual Meeting, 2006<br>Title: *The Importance of CODES Involvement in the Connecticut TRCC*</div> |

Abstract: The core focus of most CODES projects is typically on the technical aspects of data linkage, analysis of linked data, and the development of State-specific applications of the data. However, building relationships with key stakeholders is an important activity that CODES staffs undertake. Involvement in the State's Traffic Records Coordinating Committee (TRCC) is a way to learn about key data systems and programs, inform constituents about the CODES project, and identify ways to collaborate with partners to meet each other's needs. Connecticut CODES staff takes an active role in the State's TRCC. Most recently, content about the CODES project was submitted as part of the TRCC's 5-Year Strategic Plan. The Connecticut CODES staff was able to highlight positive attributes of the program, identify deficiencies and barriers to progress, and recommend ways in which these issues can be addressed. The Strategic Plan identified CODES as one of the top priorities for State traffic records improvements. The Strategic Plan was only recently submitted. The needs of CODES, as highlighted in the Strategic Plan, will help advocate for the program both within the Connecticut Department of Public Health (DPH) and statewide.

| | |
|---|---|
| Contact Person and Number | Justin Peng, 860-509-7774<br>Marian Storch, 860-509-7791 |
| Population Targeted | Administrators with decision-making authority |
| Traffic Safety Issue Targeted | **Data Quality and Content**<br>Expansion of the State crash reporting form and enhancements to the electronic database.<br>**Program Support**<br>• Increase human resources to sufficiently staff the project.<br>• Institutionalization of the project in the agency.<br>Collaboration with internal and external constituents. |
| Targeted Audience | Connecticut TRCC members, Connecticut Department of Transportation Traffic Records Section and Transportation Safety Division, Connecticut Department of Public Health Administration. |
| Methodology | • Provide updates at TRCC meetings and promote the CODES project and uses of the data. Submit content about CODES to be included in the Connecticut TRCC's 5-year Strategic Plan. A "Combined STRAW Model" was used to identify attributes and deficiencies for each project area, and provide recommendations for improvement. |
| Results | **Meetings**:<br>TRCC meetings used as an opportunity to:<br>• Gain insight about crash data from DOT data analyst.<br>• Promote future collaboration with University of Connecticut traffic researchers.<br>Build and maintain support from TRCC members who, if needed, can advocate on behalf of the project.<br><br>**Strategic Plan**:<br>• CODES was identified as a priority area for the TRCC.<br>• **Attributes**: CODES generates linked motor vehicle crash and injury outcome data; generates population-based outcome information to better characterize crashes and their associated costs; uses linked crash and injury data to guide initiatives surrounding motor vehicle and pedestrian safety in Connecticut geared toward reducing crash-related injuries and deaths.<br>• **Barriers**: Staffing challenges have affected the progress of the CODES Project; Connecticut must purchase inpatient and ED data from a private entity, and delays have been faced in obtaining more recent (2000-2004) data for linkage and analysis. |

| | |
|---|---|
| Results (cont'd) | • **Recommendations**: There are several ways in which the Connecticut CODES project can be enhanced, including:<br>   • Increasing program capacity through additional staffing;<br>   • Enhancing the crash data system (quality and content);<br>   • Expanding the crash reporting form to include fields important to CODES (e.g., gender for passengers, if injured person was transported, increased detail on the type of occupant protection used);<br>   • Linking crash and hospital data sources to EMS data (once available);<br>   • Expanding linkage efforts to include other data sources such as mortality, trauma registry, and medical examiner data; and<br>   • Linking ancillary data sources (e.g., DMV driver and vehicle data) to provide additional data for linkage and/or analysis.<br>CODES staff were recruited to be part of a subcommittee working on improvements to the crash reporting system (quality and content), including conforming to Model Minimum Uniform Crash Criteria Standards. |
| Dissemination Formats | **Meetings**: Reinforced the value of CODES data, but also discussed barriers to progress (primarily, delayed receipt of the hospital data). The TRCC facilitator included updates about CODES on nearly every meeting agenda.<br>**Strategic Plan**: Provided written content for the Strategic Plan for CODES. Also presented information, informally, at TRCC meetings. |
| Impact, Follow-Up, or Later Development on the Targeted Issue | The Strategic Plan was only recently submitted. The needs of CODES, as highlighted in the Strategic Plan, will help advocate for the program both within DPH and statewide. CODES staff will continue to use the TRCC meetings as a way to network with potential collaborators, data owners and/or data users, and to build support among the traffic records and safety community. The TRCC is submitting an application for "State Traffic Safety Information System Improvement" funds. If funded, several of the proposed projects will improve the quality and content of traffic records data, which will benefit all traffic data system users, including the Connecticut CODES project. |

# Connecticut CODES

**Connecticut Department of Public Health**

State Application Presented at CODES Annual Meeting, 2007

Title: *Providing and Improving Usefulness of CODES Data for CODES Advisory Board Members and Using CODES Data to Improve Connecticut's Child Passenger Safety Law*

Abstract: For 2006–2007, Connecticut Crash Outcome Data Evaluation System primarily focused on data linkage activities and the formal establishment of Connecticut CODES advisory board. Connecticut CODES has built strong relationships with key CODES stakeholders through constant communications in the past, but an advisory board was not formally established till 2006. An analysis using 1999 linked data was presented to the advisory board, focusing on motor vehicle occupant protection system or restraint use. Data presented included characteristics of people involved in crashes, characteristics of people not using protection systems, and outcomes of protection system use. The Director of Connecticut Safe Kids, a member of the advisory board, expressed an interest in focusing on child passenger safety (CPS). CODES Staff met with the Director of Connecticut Safe Kids to identify specific analyses of CODES data and the data presentation formats that will be most effective in convincing legislators to change the CPS law. Connecticut child passenger safety advocates, including the State and local Safe Kids Coalitions, plan to advocate for legislative changes that will bring Connecticut's law in closer compliance with national "best practice" recommendations.

| | |
|---|---|
| Contact Person and Number | Justin Peng, 860-509-7774<br>Marian Storch, 860-509-7791 |
| Population/ Problem Targeted | Injuries to children in passenger motor vehicle crashes |
| Targeted Issue | Improve Connecticut child passenger safety (CPS) laws. Connecticut's current law, effective in 2005, requires children age 6 or younger and children up to 60 pounds to ride in a child restraint system. Connecticut child passenger safety advocates, including the State and local Safe Kids Coalitions, are planning to advocate for legislative changes that will bring Connecticut's law in closer compliance with national "best practice" recommendations. |
| Requesting Office or Targeted Audience | Requesting Offices: Connecticut Safe Kids Coalition and the Connecticut Department of Public Health Injury Prevention Program.<br>Targeted Audiences: State and local Safe Kids Coalition members and other injury prevention advocates, Connecticut CODES advisory board members.<br>Legislators (reached through the State and local Safe Kids members and other advocates.) |
| Data Used | Linked/imputed data from 1999 crash, inpatient, and emergency department visit were used for this application. Only data from 1999 were available at the time of application development, and updated analyses using more recent linked data will be provided in the future. The crash data contain all crash records statewide and have most of the data elements needed for this application. The inpatient and ED visit data contain data from 30 out of 31 acute care hospitals in Connecticut and have all needed data elements for this application. |
| Methodology and Analytical Results | Data presented to the advisory board include characteristics of people involved in crashes, characteristics of people not using protection systems, and outcomes of protection system use. Further analyses focused on the outcomes of using versus not using child safety seats among children less than 5 years of age. The results found that any restraint use significantly reduced the likelihood of injury, likelihood of seeking medical care, medical charges, and length of stays. The results also found that the use of child restraint system has a further effect, although not statistically significant, on all above mentioned outcomes when compared to the use of seat belt without child restrain system. A brief explanation of the linkage imputation techniques was discussed to let the audience appreciate the complicity involved in the linkage process, but the explanation was kept simple to avoid confusion. |

| Connecticut CODES | |
|---|---|
| **Connecticut Department of Public Health** | |
| State Application Presented at CODES Annual Meeting, 2007 | |
| Title: *Providing and Improving Usefulness of CODES Data for CODES Advisory Board Members and Using CODES Data to Improve Connecticut's Child Passenger Safety Law* | |
| | |
| Dissemination Formats | A PowerPoint presentation and handouts were used to communicate the data-driven message at the quarterly Connecticut CODES advisory board meeting. The presentation started with an overview of linkage and analytical methods. It then moved into describing the characteristics of people involved in crashes and the characteristics of people not using protection systems. Finally, it ended with describing the medical outcomes of protection system use. Most results were presented in graphical formats with data values appearing on the graphs. This format was effective in providing visualizations and getting the message across. |
| Impact, Follow-Up, or Later Development on the Targeted Issue | Connecticut CODES staff met with the director of Connecticut Safe Kids, a member of the advisory board, to identify specific analyses of CODES data and the data presentation formats that will be most effective in convincing legislators to change the CPS law. It was agreed that Connecticut CODES will provide further analyses to Safe Kids Coalitions using most recent linked data and focusing on trends of child restraint system use, CRS use by age, injury severity, level of care received, average medical charges, and length of stays on statewide and county levels. Based on past experience, it is most useful for Safe Kids Coalitions to receive results in tabular format with key points highlighted. This will allow State and local level Safe Kids Coalitions the flexibility to customize the formats used to present data as needed (i.e., fact sheets, constituent letter or post card campaigns, information packets for legislators and testimony at public hearings). The Connecticut Safe Kids is now starting to plan for introduction of legislation during Connecticut's Spring 2008 legislative session. Police crash reports were the only State-Specific data set available to support previous legislative initiatives. The linked data provided by the Connecticut CODES Project to the Connecticut and local Safe Kids Coalitions will provide more comprehensive information and help them better advocate with legislators for changes to the CPS law. |

## Delaware CODES

**Delaware Office of Emergency Medical Services**

www.dhss.delaware.gov/dhss/dph/EMS/codeshome.html

State Application Presented at CODES Annual Meeting, 2006

### Title: *Support for Passenger Limitation on 16- and 17-Year-Old Drivers*

Abstract: Motor vehicle crashes are the leading cause of death among teens. Because of the high crash rate for teens, Delaware implemented a graduated driver licensing (GDL) program on July 1, 1999. One of the high-risk factors for 16- and 17-year-old drivers involved in crashes is carrying passengers. Amending the Delaware GDL system to include a limit of ***one passenger*** is expected to further reduce crashes among these drivers.

| | |
|---|---|
| Contact Person and Number | Laurie Lin, 302-378-5205 or 302-744-5400 |
| Population Targeted | Teen Driver |
| Traffic Safety Issue Targeted | Graduated Driver Licensing |
| Requesting Office or Targeted Audience | State Highway Safety Office<br>State legislative liaisons |
| Description of linked data | Data Sources: Delaware crash reports, EMS, and hospital discharge data years used: Linked data 1998-2003 |
| Analytical Results | Descriptive statistics and rate ratios (RRs) were used to compare the crash outcomes between 16- and 17-year-old drivers with and without passengers. The majority of passengers riding in cars driven by 16- and 17-year-olds involved in crashes were between 15 and 19 years old. Compared to driving alone or with one teenage passenger, 16- and 17-year-old drivers with more than one teenage passenger were:<br>(a) 1.65 times more likely to involve speeding;<br>(b) 2.19 times more likely to involve drinking and driving;<br>(c) 28 percent less likely to wear seat belts; and<br>(d) When hospitalized, spent one more day in the hospital and at an average charge of $500 more for the hospital charges. |
| Dissemination Formats | A fact sheet highlighting the information of data supports passenger limitation for 16- and 17-year-old drivers. |
| Impact, Follow-Up, or Later Development on the Targeted Issue | After reviewing the fact sheet by CODES stakeholders and highway safety offices, these stakeholders forwarded the fact sheet to their legislative liaisons to support upgrading the GDL law. Rep. John F. Van Sant, the sponsor of the bill, when he was interviewed by the *News Journal* said he believed the bill stands a good chance of passage after the General Assembly returns on May 30. |
| Comments | Delaware's GLD program has been successful and recent legislation helps address concerns with cell phone distractions. This component of House Bill 256 w/HA 5 provides additional safeguards for Delaware's driving youth by increasing the minimum age for a Level 1 Learner's Permit—also known as the graduated driver's license—from 15 years, 10 months to 16 years old. It further makes the program 18 months long, in two segments of 9 months each, adopting the restrictions of the current 1-year program, except that instead of allowing two passengers during the period of restriction, the bill allows only one passenger. In addition, the bill would require the permit holder not only to carry the permit while driving, but also require the permit holder to post a sticker or other identifying marker on the car being driven by the permit holder, indicating that he or she is operating the vehicle under a graduated driver's license. In addition, the bill requires the permit holder and all passengers under 18 to wear a seat belt or be secured in a child safety seat when the vehicle is in motion. |

| Delaware CODES | |
|---|---|
| **Delaware Office of Emergency Medical Services** | |
| www.dhss.delaware.gov/dhss/dph/EMS/codeshome.html | |
| State Application Presented at CODES Annual Meeting, 2007 | |
| Title: *Aggressive Driving Study* | |
| Abstract: Aggressive driving has become increasingly the focus of study for traffic safety because it is an unsafe driving behavior. In 2003, aggressive driving-related crashes contributed to 53 percent of all fatal crashes in Delaware. The purpose of this study is to analyze/identify aggressive driving behavior—and its consequences in terms of crashes, injuries, and fatalities— in order to improve the safety of Delaware's roadways. | |
| Contact Person and Number | Laurie Lin, 302-378-5205 or 302-744-5400 |
| Population/ Problem Targeted | Aggressive drivers |
| Targeted Issue | Reduce injuries from aggressive driving-related crashes |
| Requesting Office or Targeted Audience | State Highway Safety Office CODES stakeholders |
| Data Used | Data Sources: Delaware crash reports, EMS, and hospital discharge data<br><br>Years used: Probabilistically linked data 1998–2003 |
| Methodology and Analytical Results | Aggressive driving is defined by one or more of the following traffic violations: speeding, failing to yield right of way, passing stop sign, disregarding traffic signal, driving left of center, passing improperly, following too close, and making an improper turn.<br><br>In 2003, 52 percent of traffic-related fatalities in Delaware were aggressive driving-related, which was higher than the 47 percent aggressive driving-related fatalities in Delaware in 1998. The top three primary contributing circumstances for fatal aggressive driving-related crashes from 1998 to 2003 were failing to yield right of way, speeding, and driving left of center.<br><br>For aggressive-driving-related crashes, the death rate per 100,000 for the 15 to 19 age group increased from 16.9 in 1998 to 25.1 in 2003. Male drivers accounted for 57 percent of drivers involved in aggressive driving-related crashes.<br><br>Average hospital charges for people who were unrestrained were larger than for those who were restrained in aggressive driving-related crashes. Unrestrained people involved in aggressive driving crashes had charges $1,000 higher for inpatients when compared to restrained people. |
| Dissemination Formats | The result of the study was presented using a PowerPoint format. Crash statistics such as percentages, averages, and rates were used to show the effects of aggressive driving. |
| Impact, Follow-Up, or Later Development on the Targeted Issue | Continue to monitor and analyze aggressive driving-related data<br><br>Support the aggressive driver public awareness campaign from Office of Highway Safety |

## Georgia CODES

**Georgia Department of Human Resources, Department of Public Health**

State Application Presented at CODES Annual Meeting, 2006

Title: *Motorcycles in Georgia*

Abstract: Two separate fact sheets about injuries due to motorcycle crashes in Georgia were developed to meet the information needs of different audiences. One fact sheet consisted of a one-page summary created to inform policy makers and the general public. The other fact sheet was a more detailed descriptive analysis on motorcycle crashes and motorcycle hospitalizations that was intended for health and safety professionals. The fact sheets included crash and hospitalization data displaying crash numbers and rates, helmet use, injury severity, and contributing factors. The hospital data set contributed information about hospitalization numbers, injury severity, total hospitalization charges, and injured body region. The 2001 linked data sets were used to evaluate how motorcycle helmets were protective against traumatic brain injury.

| | |
|---|---|
| Contact Person and Number | Denise Yeager, 404-657-4776 |
| Population Targeted | Motorcycle operators and riders |
| Traffic Safety Issue Targeted | Support helmet use on motorcycles and reduce the number of deaths and severity of injuries |
| Requesting Office or Targeted Audience | Analysis requested by the CODES Board of Directors through its chair, the director of Governor's Office of Highway Safety. Other targeted audience include the general population, legislators, highway safety office/plan, injury prevention section, and public health/academic professionals. |
| Description of Linked Data | Data sources included the 2001 crash and hospital inpatient discharge data sets. The 2001 CODES linked data used the probability methods current at the time. The data quality of the crash and hospital data set was complete and validated. Since date of birth is only available from drivers in the crash data set, there was a limitation with the data among the motorcycle rider passengers. |
| Analytical Results | Linked data from crash records and hospitalizations for motorcycle related injuries were analyzed for median and average hospital charges and length of hospital stay. Primary payer sources, age-specific injury hospitalization rate, and primary injured body region were also analyzed. The odds ratio of a person having TBI without helmet usage compared to those who wore helmets was calculated to obtain TBI risk reduction of helmet usage among motorcycle riders. At least 12 percent of motorcycle riders in Georgia involved in crashes were not wearing helmets. Helmets are estimated to be 37 percent effective in preventing motorcycle fatalities and 67 percent effective in preventing brain injuries. In 2003, helmets saved an estimated 53 lives in Georgia. Significant findings were that riders wearing helmets were 30 percent less likely to have TBI than riders not wearing helmets. |
| Dissemination Formats | Two fact sheets were developed. One fact sheet consisted of a one-page summary with bullets and two pie charts to inform policy makers and the general public about motorcycle crashes. The other fact sheet was a front and back detailed descriptive analysis on motorcycle crashes and motorcycle hospitalizations also using bullets, charts, and graphs. These fact sheets were presented and distributed during a CODES board meeting and put on the injury prevention Web site. One is currently posted at http://health.state.ga.us/pdfs/prevention/MotorycycleFactSheet-Professionals.pdf |
| Impact, Follow-Up, or Later Development on the Targeted Issue | Each year there are efforts to repeal the Georgia law (OCGA 40-6-315) that requires all motorcycle riders to wear a helmet. Efforts to repeal Georgia's motorcycle helmet law during the 2005-06 legislative session were unsuccessful.<br><br>The fact sheets were discussed during a meeting of the Governor's Strategic Highway Safety Plan Risk, Analysis and Evaluation Team, and a summary of data related to motorcycles is included in the Strategic Highway Safety Plan emphasis area. |

# Georgia CODES

**Georgia Department of Human Resources, Department of Public Health**

State Application Presented at CODES Annual Meeting, 2007

## Title: *Effects of Pickup Trucks Seat Belt Exemption*

Abstract: Currently, pickup trucks are exempt from the primary seat belt law in Georgia. Georgia CODES provided data to support the House bill to eliminate the current exemption in Georgia code pertaining to the use of seat belts in pickup trucks. Pickup truck occupants with no restraints were 2.7 times more likely to be admitted to the hospital than those who wore seat belts. The average total hospitalization cost for an unrestrained pickup truck occupant is $44,000 compared to $25,000 for restrained pickup truck occupants. Child occupants in pickup trucks had the highest proportion of unrestraint use in all passenger vehicles.

| | |
|---|---|
| Contact Person and Number | Denise Yeager, 404-657-4776 |
| Population/ Problem Targeted | Pickup truck crash injury rates statewide. |
| Targeted Issue | Seat belts as worn in both the front and back seat of all passenger vehicles including pickup trucks. |
| Requesting Office or Targeted Audience | Analysis requested by the CODES board of directors through its chair, the director of Governor's Office of Highway Safety. Other targeted audiences include the Division of Public Health, legislators, highway safety office/plan, injury prevention section. |
| Data Used | The years of imputed linked data used were 2001 and 2002. The 2001 linked data included crash, EMS, and hospital Inpatient data only. The 2002 linked data included crash, EMS, and hospital emergency department and inpatient discharge data. Missing values were not imputed. The hospital and crash data were in most cases complete and representative with the full crash data using vehicle type. |
| Methodology and Analytical Results | Linked data from crash records and inpatient hospitalizations for pickup truck related injuries were analyzed for median and average hospital charges, length of hospital stay, and primary payer sources. The odds ratio of a person that is admitted to the hospital without a seat belt compared to those who used seat belts was calculated. Significant findings were that occupants not wearing seat belts were 2.7 times more likely to be admitted to the hospital than those using seat belts. |
| Dissemination Formats | Information was included in the internal legislative impact statement for the Division of Public Health. A narrative was included in this statement and included a statement that CODES data was used. A fact sheet was developed and is viewable online at http://www.legis.state.ga.us/legis/2005_06/house/ Committees/motorVehicles/motorAgendas/MVhand2.pdf |
| Impact, Follow-Up, or Later Development on the Targeted Issue | Highway Safety Office and Division of Public Health supported House Bill 608 and Senate Bill 86 to eliminate the exception. The House did not vote on HB 608 and SB 86 passed the Senate. |

# Illinois CODES

**Illinois Department of Transportation, Division of Traffic Safety**

http://www.dot.il.gov/trafficsafety/IRTCC.html

State Application Presented at CODES Annual Meeting, 2007

## Title: *Inpatient Hospital Charges and Utilization Patterns of Crash Victims in Illinois*

Abstract:  Illinois has linked 2002 crash and augmented hospital inpatient discharge data to obtain information on traffic-related injury cases discharged from Illinois hospitals. Specifically, the data contains information on the number of discharges, average charges, average length of stay, primary injuries, types of crash controlling for demographics (age and gender), region, expected payment source and discharge status, vehicle type (passenger car, light truck, motorcycle), occupant restrained used, and several factors. Preliminary findings have been presented to the Illinois CODES Advisory Group and made available to the Illinois Traffic Records Coordinating Committee.

| | |
|---|---|
| Contacts | Mehdi Nassirpour, 217-785-8905<br>Susan Fitzpatrick, 217-785-0281<br>Kristopher Boyer, 217-785-3041 |
| Population/ Problem Targeted | All individuals involved in crashes in 2002 are included in this study of hospital inpatient charges and utilization patterns, and the health consequences of unrestrained drivers and passengers. |
| Targeted Issue | Develop support for safety legislation (primary belt for back seat passengers and helmet use). |
| Requesting Office or Targeted Audience | The Illinois CODES project was initiated by the Division of Traffic Safety at IDOT in cooperation with the Illinois Department of Public Health. Preliminary results of the linked crash and hospital discharge data were presented to the Illinois CODES Advisory Group, ITRCC, and Highway Safety Planning committee at IDOT. |
| Data Used | Illinois linked 2002 hospital inpatient discharge records with 2002 trauma discharge records in order to replace missing E-codes on hospital inpatient records with E-codes found on probabilistically linked trauma records. Illinois linked 2002 person level crash records with the 2002 augmented hospital inpatient discharge records using CODES2000's probabilistic record linkage procedures to create five separate multiply imputed data sets of linked records. |
| Methodology and Analytical Results | The imputed linkage results served as the input data set to SAS' PROC MI to create multiple imputations replacing missing values with estimated values. SAS' PROC MIANALYZE was used to evaluate means and standard errors from the PROC MI results. |
| Dissemination Formats | The following reports were presented to the CODES Advisory Committee and made available to the Illinois Traffic Records Coordinating Committee to illustrate the linkage process and overall results:<br>· Descriptive statistics (e.g., number, mean, median) for charges and length of stay from four data sources: the source hospital inpatient data, the augmented hospital inpatient data, high probability linkages and imputed linkages.<br>· Descriptive statistics for charges from the four data sources by three E-code response categories (motor vehicle traffic crash, other than motor vehicle traffic crash, and missing). Preliminary results were presented and discussed at the Illinois CODES Advisory Group meeting. These results included estimated average charges and length of stay by hospitalization variables (i.e., discharge status, site of first injury, payer type) and by crash variables (i.e., belt status, collision type, gender, helmet usage, intersection related, KABCO, number of vehicles, occupant position, region, road type, severity, time of day, and weekday/weekend). |
| Impact, follow-up, or later development | The presentation generated much useful discussion and suggestions for improvements in the linkage and reporting process. It also sparked increased interest in the CODES Project as evidenced by prompt follow-up to questions and suggestions. |

# Indiana CODES

**Indiana Department of Homeland Security**

http://cats.ecn.purdue.edu/CODES.aspx

State Application Presented at CODES Annual Meeting, 2007

## Title: *Effect of Seat Belt Usage on Hospital Charges by Vehicle Type*

Abstract: Indiana has made significant strides in improving Statewide seat belt usage rates, with rates increasing from 62 percent in 2000 to around 84 percent in December 2006. New legislation that mandated front seat passengers to buckle up was largely responsible for this increase. Pickup trucks, however, were excluded from the law and usage rates for pickup trucks remained flat in Indiana from 2003–2006. As a new bill to close the exclusion for pickup trucks was being debated in the Indiana State House and Senate, CODES data were used in an attempt to show the financial and physical consequences of not wearing a seat belt. Crash, EMS, and hospital data for the years of 2003-2005 were linked using NHTSA's CODES2000 software.

The CODES data showed that on a yearly basis in Indiana, there were around 330,000 crash occupants, 130,000 EMS patients, and 550,000 in- and out-patients (filtered to include only patients that had an injury). Occupants of cars and pickup trucks that were linked to hospital data had their charges and length of stay cross tabulated as a function of their restraint usage status. Restrained occupants showed between 40 percent and 50 percent lower charges than nonrestrained occupants. The results were displayed in a fact sheet and presented to the Indiana TRCC and eventually passed on to legislators. House Bill 1237 passed the Senate 31 to 16 and the House 58 to 42 and was signed into law. This bill ended the exception in Indiana's mandatory seat belt law, and now requires all occupants, including backseat passengers in all vehicles, to buckle up.

| | |
|---|---|
| Contact Person and Number | Jose Eduardo Thomaz, Data Warehousing Administrator, Center for Road Safety, Purdue University, 765-406-1350, jthomaz@ecn.purdue.edu |
| Population/ Problem Targeted | Pickup trucks occupants were exempted from Indiana's seat belt law. The study was targeted to legislators to demonstrate the effects of that legal exemption on hospitalization charges and length of stay. |
| Targeted Issue | Indiana seat belt laws only required seat belt use by front seat occupants, and exempted more than 1.4 million vehicles registered as pickup trucks (as well as vans and SUVs whose owners have paid an additional fee at the Bureau of Motor Vehicles to get a truck plate). The CODES team hoped to demonstrate the difference in hospital charges and length of stay of restrained and nonrestrained occupants, for both classes of vehicles, to support changing the legislation to close the pickup truck exception. |
| Requesting Office or Targeted Audience | The preliminary results of the study were presented to the TRCC board, at the request of the Indiana Criminal Justice Institute, Traffic Safety Division. A fact sheet was then passed on to the legislators. |
| Data Used | Crash, EMS and hospital data for the years of 2003–2005 were linked using NHTSA's CODES2000 software. On a yearly basis, there were around 330,000 crash occupants, 130,000 EMS patients and 550,000 in- and out-patients (filtered to include only patients that had an injury). Overall, between 4,500 and 5,000 vehicle occupants involved in crashes were successfully linked to EMS and hospital data. Another 21,000 vehicle occupants were linked to hospital data, but not to an EMS transport. |
| Methodology and Analytical Results | The analysis data underwent imputation during the linkage and the modeling process. The model was adjusted to accommodate Indiana data reliability levels, giving more emphasis on hospital data as covariates whenever possible, since hospital data have the highest reliability. |
| Dissemination Formats | The CODES analysis was summarized in a fact sheet with tables and charts that broke down hospital charges and lengths of stay by vehicle type, restraint use, and injury severity. This analysis made clear that charges for nonrestrained occupants were almost double those of restrained ones. These tables were discussed during the CODES/TRCC monthly meeting and quickly passed on to the Indiana senate members debating House Bill 1237. |
| Impact, Follow-Up, or Later Development on the Targeted Issue | House Bill 1237 passed the Senate 31 to 16 and the House 58 to 42 and was signed into law. This bill closed the loophole in Indiana's mandatory seat belt law, and now requires all occupants, including back-seat passengers in all vehicles, to buckle up. |

# Iowa CODES

**Iowa Department of Public Health/Center for Vital Records and Health Statistics**

http://www.idph.state.ia.us/apl/codes.asp

State Application Presented at CODES Annual Meeting, 2006

## Title: *Motor Vehicle Crash Related Traumatic Brain Injuries in Iowa (2001-2003)*

Abstract: Analyses were conducted to demonstrate the magnitudes of motor vehicle crash (MVC) injuries in Iowa, especially the traumatic brain injuries caused by MVC, to compare the differences between TBI and other injury type by risk factors such as vehicle type, crash type, safety equipment, and financial consequences.

| | |
|---|---|
| Contact Person and Number | Suning Cao, 515-281-3983<br>Scott Falb, 515-237-3154 |
| Population/ Problem Targeted | Statewide MVC injured people who were hospitalized (2001–2004). Those who were linked to crash records (2001–2003) were analyzed by vehicle type, crash type, restraint use, age group, and alcohol involvement by comparing injury type: TBI and non-TBI. |
| Issue Targeted | Vehicle type, crash type, restraint use, younger riders, impaired drivers, and their interrelations. |
| Requesting Office or Targeted Audience | As the co-chairperson of Statewide Traffic Records Coordinating Committee (STRCC), Scott Falb requested this study and we presented it together to the full member meeting of STRCC. Also used for occupant restraint task force group within STRCC and Iowa's Safety Management Systems; Presented at Prevention of Disability Council meeting. Published at IDPH CODES Web site to educate the public. |
| Data Used | Three-year imputed crash-inpatient links (4,551 records, 2001–2003) were studied, including occupants of passenger cars, van, pickups or SUVs, pedestrians, bikers, and unknown vehicle type riders, who were hospitalized in Iowa. |

## Methodology and Analytical Results

First, using Iowa hospital inpatient discharge data (2001–2004), to demonstrate the estimated medical costs of MVC-related TBI versus other injury type; children and younger riders versus adults:

- Each year about 2,000 people injured in MVC are hospitalized;
  - 27 percent of the hospitalized sustained TBI;
  - Total medical costs for MVC TBI patients account for 60 percent of all MVC-related injuries costs.

Second, an analysis compared linked cases (4,550 records, 2001–2003) by variables of vehicle type, crash type, impaired drivers and restraint use, etc. in two groups: TBI (n=1,292, 28%) and other injury (3,258, 72%). The odds ratios were calculated. Results:

- Not wearing a seat belt or helmet is a significant risk factor in sustaining TBI for MVC victims.
- Unrestrained pickup/SUV occupants were twice as likely to sustain TBI as the restrained, while unrestrained passenger car occupants were 1.5 times more likely to suffer TBI than restrained occupants.
  - 24 percent of passenger car occupants hospitalized were not wearing seat belts; 35 percent of pickup/SUV riders did not wear a seat belt

The TBI rate between pickup/SUV and passenger cars was not statistically different.

- Injured bicyclists were the most vulnerable to sustain TBI once involved in MVC: 54 percent of bicyclists were children between the age of 5 and 14.
- Injured pedestrians were the second most vulnerable group in suffering TBI. Pedestrians were 1.8 times more likely to sustain TBI than passenger car riders. 33 percent of the pedestrians sustaining TBI were under the age of 15.
- Impaired-drivers accounted for 15 percent of all hospitalized MVC injuries and they were 1.4 times more likely to sustain TBI than nonimpaired drivers.
- 44 percent of hospitalized MVC injuries occurred in single-vehicles crashes in Iowa, and single vehicle crashes were 1.3 times more likely to cause TBI than multiple-vehicle crashes.

TBI patients were 5 times more likely to die at the hospital than non-TBI patients.

Third, cost estimations demonstrated that:

- Estimated medical costs in Iowa for a TBI patient was $120,060 (median value, 2003) versus $11,924 for a non-TBI patients.

Estimated other costs for a TBI patient was $127,654 versus $61,728 for a non-TBI patient.

| | |
|---|---|
| Dissemination Formats | Presented the analysis to STRCC meeting and to the Prevention of Disability Council Meeting |
| Impact, Follow-up, or Later Development | The Iowa task groups are still finalizing all data resources. These data are included as part of the proposed legislation for rear passenger safety-belt mandates. |

# Iowa CODES

**Iowa Department of Public Health/Center for Vital Records and Health Statistics**

State Application, 2007

http://www.idph.state.ia.us/apl/codes.asp

## Title: *Iowa Motor Vehicle Crash Injury Facts Among Emergency Department Visits: A Profile of Unbelted Occupants of Passenger Cars and Vans/SUVs (2003)*

Abstract: Analyses were conducted to demonstrate the magnitude of MVC injured people who visited Iowa emergency departments (2003), compare the differences among them by discharge status and seat belt use, and analyze the impact of non-belt use on ED patients' outcomes and medical charges.

| | |
|---|---|
| Contact Person and Number | Suning Cao, 515-281-3983<br>Scott Falb, 515-237-3154 |
| Population/ Problem Targeted | Statewide MVC injured people who visited emergency department and were linked to crash records. |
| Targeted Issue | Seat belt use and its association with age, gender, alcohol use or drug influence, ejection, air bag, crash location, seating position, collision type, and vehicle damage. |
| Requesting Office or Targeted Audience | As the co-chairperson of Statewide Traffic Records Coordinating Committee (STRCC), Scott Falb requested this study and we presented it together to the full member meeting of STRCC.<br><br>Used for occupant restraint task force group within Statewide Traffic Records Coordinating Committee (STRCC) and Iowa's Safety Management Systems.<br>Published at IDPH CODES Web site to educate the public. |
| Data Used | 10,475 (95.5% of total crash and ED linked/imputed data for 2003) occupants of passenger cars, van, pickups or SUV, 16 or older, who were treated at ED's, were studied. Seat belt analysis excluded the cases where seat belt use was unknown (28% of ED links). |
| Methodology and Analytical Results | First, crash-ED linked patients were compared in the demographic, TBI, medical charges, safety devices, vehicle, collision, and environment variables by three discharge status:<br>   1. Discharged from ED;<br>   2. Hospital Admission from ED;<br>   3. Died at ED.<br><br>Second, an analysis was narrowed down to the non-missing belt information cases by belted and nonbelted groups for comparisons.<br><br>The result was that unbelted occupants are more likely to become inpatients and have a higher percentage of TBI than belted occupants. |
| Dissemination Formats | Distributed the fact sheet during STRCC meeting to show that by adding ED links the imputed linked cases increased to about 12,000 annually from about 2,000 inpatient links only. |
| Impact, Follow-Up, or Later Development on the Targeted Issue | Task groups are still finalizing all data resources. These data are included as part of the proposed legislation for rear passenger safety-belt mandates. |

| | |
|---|---|
| **Iowa CODES**<br>**Iowa Department of Public Health/Center for Vital Records and Health Statistics**<br>http://www.idph.state.ia.us/apl/codes.asp<br>State Application Presented at CODES Annual Meeting, 2007<br>Title: *Motorcycle Quick Facts in Iowa – 2001-2005* | |
| Abstract: To demonstrate the growing crash and injury rates among motorcycle riders in Iowa from 2001–2005, the five-year's crash rates per 1,000 motorcyclist licensed drivers — including fatality and injury rates — were calculated. The goal was to show the growing number of injured riders (especially for people age 45 to 64) by using inpatient data and the cost impact on government payer sources. Using 2003 crash-ED-inpatient-linked data to demonstrate the association between helmet use and brain injury and higher costs for the unhelmeted riders. | |
| Contact Person and Number | Suning Cao, 515-281-3983<br>Scott Falb, 515-237-3154 |
| Population/ Problem Targeted | Statewide motorcycle injured people, those who visited ED and those who were hospitalized, with regard to helmet use and brain injury. |
| Targeted Issue | The high crash rate among young riders (age 15 to 24).<br>The increased crash rate among riders in the 35–to-54 age group.<br>Helmet protection from brain injury and a lower helmet use in Iowa. |
| Requesting Office or Targeted Audience | As the cochairperson of Statewide Traffic Records Coordinating Committee, Scott Falb requested this study and we presented it together to the full member meeting of STRCC.<br><br>Published at IDPH CODES Web site to educate the public. |
| Data Used | One year crash-ED-inpatient imputed links (2003). For the fact sheet, only the first imputed linked data set was analyzed, where cases without helmet use information (11% of the total inpatient links) were excluded. |
| Methodology and Analytical Results | First, using Iowa crash data from 2001–2005, motorcycle crash rate, fatality rate and injury rate were calculated for each year to demonstrate the increasing crash and injury rates in the State. By analyzing the crash rates by age groups, the results showed that younger motorcyclists, age 15 to 24, were 7.5 times as likely to have a crash than riders older than age 24 on the five years average.<br><br>Second, using 2003–2005 Outpatient ED data, 2,510 motorcycle-related ED visits were analyzed by age group, medical charges, and dispositions.<br><br>Third, using 2001–2005 inpatient data to show the increasing growth trend — a 9 percent increase from 2001 to 2005—in Iowa's motorcycle-related hospitalizations. Among them, the age groups 45 to 54 and 55 to 64 had the highest rate increases, 76 percent and 190 percent, respectively.<br><br>By using 2003 CODES linked data to demonstrate the strong link between not wearing a helmet and sustaining severe TBI, CODES data show that while 24 percent of unhelmeted riders had TBI, none of the helmeted riders suffered severe TBI and that there were higher hospital charges for unhelmeted riders (2.3 times higher than for helmeted riders).<br><br>Conclusion: From the view of public health and safety, there is repeated evidence that helmets prevent head injury and reduce the likelihood of a crash fatality. |
| Dissemination Formats | The fact sheet was distributed at the Iowa Motorcycle Safety Forum 2007. |
| Impact, Follow-Up, or Later Development on the Targeted Issue | These data results have been included in Iowa's Safety Management Systems within Traffic Safety. The data results are being used to support legislation and Iowa's goal of mandating helmet use. |

## Kentucky CODES
### Kentucky Injury Prevention and Research Center
www.kiprc.uky.edu/projects/CODES

State Application Presented at CODES Annual Meeting, 2006

## Title: *Economic Costs of Low Seat belt Usage in Kentucky*

Abstract: NHTSA has published estimates of the effectiveness of seat belts in preventing fatal and nonfatal injuries. The estimated effectiveness varies depending on the type of vehicle (passenger car versus light truck), the type of seat belt used (lap belt only versus lap and shoulder belt, and the occupant's seating position. Linked collision report and hospital discharge data were used to estimate Kentucky's overall seat belt effectiveness for preventing moderate-to-critical injury at 55 percent. This number was then used to calculate medical cost savings for Kentucky's Medicaid program and other payers that would result from enactment of a primary enforcement law. Particular attention was given to traumatic brain injuries and spinal cord injuries because they often result in medical costs for multiple years after the original injury.

| | |
|---|---|
| Contact Person and Number | Michael Singleton, 859-257-5809, msingle@email.uky.edu |
| Population Targeted | Motor vehicle occupants |
| Traffic Safety Issue Targeted | Support for primary belt use |
| Requesting Office or Targeted Audience | Requested by the Kentucky Transportation Cabinet, Dept of Highway Safety. Target audience also includes governor and State legislators. |
| Description of linked data | The primary data sources were the State hospital discharge databases for 2000–2004. The CODES database, with multiply imputed links for 2003, was used to generate a Kentucky-specific estimate of the effectiveness of seat belt usage in preventing injuries and fatalities, which varies depending on seating position, vehicle type, and seat belt type. |
| Analytical Results | It was determined that with the passage of a primary seat belt law in 2006, 57 lives could be saved and 307 hospitalizations could be prevented annually, an overall savings of at least $118 million in direct medical costs over the ten-year period from 2006 to 2015. Of this total, $34 million could be saved in total direct medical costs to Medicare alone. In addition, there would be at least $68 million saved in medical costs to commercial insurers, $23.3 million to Medicare, $3.3 million to Worker's Compensation, and $11.2 million to other sources. |
| Dissemination Formats | The Kentucky Transportation Cabinet provided copies of the report to all legislators. Many legislators, as well as the Governor, used the information regarding the number of lives saved during the debate and press conferences. The Transportation Cabinet led 12 "Saved by the Belt" rallies across the State and the information from the report was used in these rallies. |
| Impact, Follow-Up, or Later Development on the Targeted Issue | The Kentucky legislature passed a primary enforcement bill in the 2006 General Assembly. The CODES report was one of a number of factors that preceded its passage. In addition to the injury prevention and Medicaid considerations, there were several million dollars of federal highway incentive funds available to the State contingent on passage of the bill. There was debate in the legislature on Medicaid cost savings, but the main debate focused on saving lives and the incentive funding for highways. |
| Notes | These savings would be a direct result of an increase in the number of Kentuckians who would wear a seat belt if a primary enforcement law were in place. States that have enacted primary enforcement legislation have experienced increases in seat belt use of as much as 18 percentage points. NHTSA has stated that the average increase is between 10 and 15 percentage points. The savings given above assume that Kentucky would experience an increase of 14 percentage points, which would move the State's usage rate from 67 percent to 80 percent, which was the national average in 2004. |

# Kentucky CODES

**Kentucky Injury Prevention and Research Center**

www.kiprc.uky.edu/projects/CODES

State Application Presented at CODES Annual Meeting, 2006

## Title: *Injuries to Booster-Age Children in Kentucky, 2000-2004*

Abstract: Kentucky passed a primary seat belt law in 2006 without passing a booster seat law, which left many children age 4 to 8 at risk, due to being improperly restrained in lap/shoulder belts only. According to NHTSA, 80 percent to 90 percent of all children in the U.S. who should be restrained in a booster seat are not. Kentucky's linked, imputed crash and hospital database was used to compare injury diagnoses between restrained 4 to 8 year old occupants and restrained occupants of other ages. The findings confirmed that among restrained occupants who were hospitalized, a principal diagnosis of abdominal injury was more common among booster-age children than other age groups. A fact sheet was created that focused on the outcomes (injury type, hospital charges, hospital length of stay, etc.) for restrained and unrestrained children 4 to 8 years old, compared with other age groups.

| | |
|---|---|
| Contact Person and Number | Mike Singleton, 859-257-5809, msingle@email.uky.edu |
| Population/ Problem Targeted | Children of booster seat age (typically between age 4 to 8 years) who should be restrained in booster seats, but usually are not. |
| Targeted Issue | Advocate for increased use of booster seats and passage of booster seat legislation to reduce injury severity among young children. |
| Requesting Office or Targeted Audience | Booster seat coalition, Legislators, Governor's Executive Committee on Highway Safety |
| Data Used | Kentucky's 2000–2004 crash data, inpatient data, and the linked, imputed data were used to do this application. The data linkages were performed using CODES2000 version 6.1. |
| Methodology and Analytical Results | Using the linked, imputed data, injuries to children ages 4 to 8 who were restrained in a child safety seat (CSS) were compared to those who were not restrained in a CSS, according to the police crash report. Those children who were restrained in a CSS were 40 percent less likely to be hospitalized than those who were not; they also had lower hospital charges, length of stay in hospital, and overall injury severity score. |
| Dissemination Formats | At the 2007 Kentucky Lifesavers Conference in April, which is sponsored annually by the Department of Transportation Safety, the findings were presented. Video clips were used to demonstrate the biomechanical and injury issues related to booster seat usage. Tables and charts based on CODES data were also presented. |
| Impact, Follow-Up, or Later Development on the Targeted Issue | The Kentucky legislature passed a booster seat bill during the 2008 General Assembly that applied to kids age 4 to 7. The law went into effect on July 15, 2008. |

# Maine CODES
**Maine Health Information Center**
www.mhic.org/CODES
State Application Presented at CODES Annual Meeting, 2006
### Title: *Maine CODES Claims Data Bank Project*

Abstract: In 2001, the Maine Legislature and Governor King approved LD 1304 An Act to Create the Maine Health Data Processing Center (MHDPC). The MHDPC collects administrative eligibility, medical, pharmacy, and dental insurance claims from all payers providing insurance to Maine residents. 2003 was the first year of data collected for this new claims data bank (DBANK).

As one of the first projects to use this new data resource, the Maine CODES project has linked 2003 CODES injured occupants to 2,463 people in the claims data bank. For these 2,463 people, a total of 200,845 service line claims tracking up to 18 months of medical care post-crash. An algorithm was developed to determine which post-crash claims were related to the crash injuries and which were pre-existing or new medical claims for each person.

Overall, the post-crash expense—on a per-person, per-month basis—increased by 140 percent with higher rates of increase for crashes involving moose, rollovers, alcohol-related, and rural locations. The average expense for an unrestrained injured occupant ($9,272) was 3.7 times a restrained occupant ($2,491).

Of the $10.5 million in crash-related injury expenses, $4.5 million (39%) were non-hospital provider claims. Among the 2,463 occupants, claims data indicated that 3 percent were hospitalized in ICU, 8 percent had an operation, 436 (18%) had a CT-Scan, and 299 (12%) had follow-up physical therapy. Furthermore, 425 (17%) of these occupants were still receiving care and treatments for their crash injuries 90 days or more after the crash.

| | |
|---|---|
| Contact Person and Number | Karl Finison, Maine Health Information Center (MHIC), 207-430-0632, kfinison@mhic.org |
| Population Targeted | All 2003 Maine CODES occupants with linked medical records linked to statewide medical insurance claims data. |
| Traffic Safety Issue Targeted | Provides new information about the long-term medical impact and expense of motor vehicle crashes. Provides new outcome metrics that incorporate information beyond the hospital setting. |
| Requesting Office or Targeted Audience | The Data Bank project was a separately funded project we developed with NHTSA support. It was supported by all Maine offices represented on the Maine Advisory group. The Maine Health Data Organization is the State office that controls the statewide claims data base that was linked and used in this analysis. Other potential audiences include the Governor, Legislators, Highway Safety and Injury Prevention Programs. |
| Description of linked data | 2003 Maine CODES crash occupants with linked medical records were linked to new statewide claims data bank. 2,643 injured occupants linked to the claims data bank. For these 2,643 people, 200,845 service line medical and pharmacy claim line services were identified for the 18-month period from January 2003 to June 2004 (most current data available). The claims data contains ICD-9 diagnosis (up to 10) to identify injury types as well as CPT, Revenue, and RX NDC coding to identify types of service. This is a first look at services provided or billed outside of the hospital setting. <br><br> Algorithms were developed to distinguish the claims that were related to injuries resulting from the crash from non-injury or pre-existing injuries using the Barell Matrix coding, which was used to assist in the process and Maine incorporated ICD-9 codes and medication for "pain." <br><br> Limitations: The Maine claims data bank is currently limited to Maine residents with private medical insurance; work continues to add Medicaid and Medicare claims. Claims paid by auto insurance are not included. |

| | Maine CODES |
|---|---|
| | **Maine Health Information Center** |
| | www.mhic.org/CODES |
| | State Application Presented at CODES Annual Meeting, 2006 |
| | Title: *Maine CODES Claims Data Bank Project* |
| Analytical Results | The resulting CODES-DBANK is a longitudinal data base. 2,463 people with 200,845 claims were linked (47,452 claims were crash-related injury claims). Analyses had the following parameters:<br><br>1. Claim Type: Claims are categorized as crash-related injury or non-crash-related (e.g., mammogram, office visit for sore throat, inpatient maternity hospitalization, pre-existing injury claims). For the 2,463 people linked, the per member per month (PMPM) claims expense increased from $328 pre-crash to $786 post-crash. $358 of the PMPM increase was due to claims identified as crash-related injury claims.<br><br>2. Time from crash: Injury claims by time interval post-crash (days, weeks, and months). 425 (17%) of the occupants were still receiving care and treatments for their crash injuries 90 days or more after the crash. 154 (6%) were still using prescription pain medication for their injury 90 days or more after the crash.<br><br>3. Injury types: Crash-related injury claims are reported by the Barell Body Region of Injury (ISRSITE2) and Barell Nature of Injury (ISRCODE). Fractures of the lower extremities, TBI, internal torso, spinal and vertebral column injuries were leading contributors to medical claims expense. 177 (7%) had a TBI injury and 143 (6%) had a lower extremity fracture.<br><br>4. Service types: Crash-related injury claims are reported by setting (e.g., inpatient, outpatient hospital, professional) and types of service (e.g., ICU, operating room, CT scans, physical therapy, office visits). Of $10.5 million in crash-related injury expenses, $4.5 million (39%) were non-hospital claims.<br>The 2,463 occupants claims indicated that 69 (3%) were hospitalized in ICU, 197 (8%) had an operation, 436 (18%) received a CT Scan, 299 (12%) had physical therapy.<br><br>5. Crash variables and selected outcome metrics: Crash-related injury claim selected indicators (e.g., claims expense, ICU, CT scans, occurrence of crash-related injury claims more than 30 days post-crash) are tabulated by selected police-reported crash variables (e.g., age of occupant, gender of occupant, seating position of occupant, vehicle type, crash type, speed limit on roadway, alcohol-related, and safety device use—such as seat restraints, air bags, and helmets). The highest increase in post-crash expenses PMPM were motorcyclists without helmets (1048%), crashes involving moose (968%), and rollovers (430%). Furthermore, 9 percent of occupants in alcohol-related crashes were hospitalized in ICU compared to 2 percent in non-alcohol related. |
| Dissemination Formats | Maine CODES creates written reports and PowerPoint presentations.<br>www.mhic.org/CODES.<br>In cooperation with the Maine Office of Data, Research, and Vital Statistics, Maine CODES creates fact sheets that are distributed via the Maine Bureau of Health Web site<br>http://www.maine.gov/dhhs/bohodr/fact_sheet_index.htm |
| Impact, Follow-Up, or Later Development on the Targeted Issue | Data is new and has not been released. Results will be reviewed with Maine CODES Advisory Committee on May 24, 2006. |

## Maine CODES
**Maine Health Information Center**
www.mhic.org/CODES
State Application Presented at CODES Annual Meeting, 2007
### Title: *Adjusting for Seat Belt Reporting: The Problem of Differential Misclassification*

Abstract: During the past 15 years, efforts to increase the use of seat belts in Maine resulted in passage of a secondary seat belt law in 1997 and a primary seat belt law in 2007. At the request of the Maine CODES Advisory Committee, CODES data has been used to support efforts to identify the benefits of seat belts, increase usage, and support legislative review. Differential misclassification of seat belt use results in inaccurate surveillance reports and exaggeration of the benefit of seat belts; the Maine project has adjusted the analyses for this problem. This presentation reviews the problem of differential misclassification of seat belt use in police crash reports and explores a method of correcting for missing data and differential misclassification using CODES linked data.

| | |
|---|---|
| Contact Person and Number | Karl Finison, Maine CODES, 207-430-0632. kfinison@mhic.org |
| Population/ Problem Targeted | Differential misclassification of seat belts in police crash reports. Uninjured or slightly injured occupants in crashes have an opportunity and often a motive to say they were wearing seat belts when they were not. This results in differential misclassification of seat belts in FARS, CODES, and other data based on police-reported information. The problem results in over-reporting of belt use rates in surveillance reports and biased estimates of the effectiveness of belts. |
| Targeted Issue | Increase understanding of seat belt usage and improved analyses |
| Requesting Office or Targeted Audience | This was a special project developed for the Charleston CODES Technical Assistance meeting in order to focus attention on a need and examine possible solutions. The methods developed have been used over the years for a variety of reporting and analyses of the seat belt issue in Maine, requested by our Maine Advisory group – primarily the Maine Bureau of Highway Safety. |
| Data Used | Maine CODES has used linked data since the 1991 crash data year for five different reports/analyses on seat belts. In this presentation, an example using 2003 linked imputed data compares risk estimates, adjusted for missing values, and adjusted for missing values and differential misclassification. |
| Methodology and Analytical Results | Different methodologies are described. Methodology and results using imputed methods to adjust for differential misclassification and results are presented, |
| Dissemination Formats | Surveillance reports, analytical reports, fact sheets, and PowerPoint presentations have been used, and the results have been provided to the Maine CODES Advisory Committee, the Maine Bureau of Health, the Governor's Office, and the State TRCC. The results have also been published on the Internet at: www.mhic.org\CODES |
| Impact, Follow-Up, or Later Development on the Targeted Issue | Fact sheets developed by Maine Bureau of Health from CODES data. CODES data used by Governor's office and various committees supporting seat belt legislation over a 15-year period. A primary seat belt law was passed in 2007. |

# Maryland CODES

**National Center for Trauma/EMS, University of Maryland-Baltimore**

http://medschool.umaryland.edu/NSCforTrauma//codes.asp

State Application Presented at CODES Annual Meeting, 2006

## Title: *CODES Data Warehouse Using Data for Strategic Planning and TRCC*

Abstract: Over the years, the CODES program in Maryland has evolved to become a defacto Data Warehouse for statewide Highway Safety Data. In addition to the three core data sets (crash, EMS, hospital) the Maryland CODES has added a significant number of ancillary data sets including licensing, registration, citation, motorcycle training, trauma registry and other that have been used individually and collectively for a variety of highway safety topics. In addition to responding to numerous data queries CODES has developed general and program-specific fact sheets, contributed data for use in the State's Highway Safety Office's annual and benchmark reports, been a regular participant in the State's Traffic Records Coordinating Committee and provided data for problem identification to each of Maryland's twenty four subdivisions. Recently, Maryland CODES has also been able to use ancillary data sets to improve blood alcohol content test reporting for NHTSA's Fatality Analysis Reporting System. During the past year, the CODES project has contributed data and expertise to the State Highway Administration to assist in the development of a statewide strategic plan that will guide the direction of traffic records and highway safety in the State for the next several years. Data supplied through CODES has been used to help the planning committee develop a list of priorities to be addressed by each program area subcommittee. Currently CODES data is being used by the Highway Safety Office to support strengthened impaired driving and young driver bills for the current legislative session and to support the 408 grant application process by providing measurable performance measures.

| | |
|---|---|
| Contact Person and Number | Timothy Kerns, University of Maryland/National Study Center for Trauma and EMS, 410-328-4244, tkerns@som.umaryland.edu |
| Population Targeted | CODES data is available as a resource for all State and local agencies as well as other stakeholders in highway safety. |
| Traffic Safety Issue Targeted | Maryland CODES is becoming established as a principal source of data and analysis for highway safety questions within the State. |
| Requesting Office or Targeted Audience | The TRCC, the Strategic Planning Committee, State Highway Administration and State legislators. |
| Description of linked data | Data sources: Crash reports, hospital discharge, citations, trauma registry, licensing, registration, motorcycle training, and other smaller data sets related to highway safety. |
| Analytical Results | Extensive analysis of various program areas including younger and older drivers, impaired driving, aggressive driving, inattentive driving. Development of a Traffic Safety Fact Book modeled on the version published by NHTSA along with program and county specific fact sheets for use within the State. |
| Dissemination Formats | Presentations given at recent strategic planning committee meetings and TRCC meetings, summary analysis reports and fact sheets. All information can be obtained through a Web site that is currently being updated to focus on CODES activities. |
| Impact, Follow-Up, or Later Development on the Targeted Issue | Data provided through CODES were used extensively in setting priority areas for strategic planning, problem identification, safety analysis, and program evaluation. |

## Maryland CODES
**National Center for Trauma/EMS, University of Maryland-Baltimore**
http://medschool.umaryland.edu/NSCforTrauma//codes.asp
State Application Presented at CODES Annual Meeting, 2007
### Title: *Comprehensive Crash Outcome Data Evaluation System*

Abstract: Data available through the Maryland CODES project are compiled into multiple annual reports covering a number of program areas that are used extensively by State highway safety partners for problem identification and program evaluation. These reports include: (1) Impact Objectives; (2) State of the State; (3) State of the County; (4) Crash-Crime clocks; (5) Program Area Over-Representation; and (6) Comprehensive State and County Fact Books. Further, CODES data are used in the annual benchmark report provided by the State's Highway Safety Office to the regional NHTSA office and in their annual report provided to the Maryland Department of Transportation. Also, CODES data have been used in the preparation of Maryland's 408 application, to support Maryland's Traffic Records Coordinating Committee, and to provide data for motorcycle and impaired driving assessments.

| | |
|---|---|
| Contact Person and Number | Timothy Kerns University of Maryland/National Study Center for Trauma and EMS, 410-328-4244, tkerns@som.umaryland.edu |
| Population/ Problem Targeted | Impaired driving, young drivers, older drivers, motorcycles, inattentive driving, aggressive driving, occupant protection, and pedestrians. |
| Targeted Issue | To provide data-driven problem identification and evaluation for a variety of highway safety topics in Maryland. |
| Requesting Office or Targeted Audience | Maryland Highway Safety Office, Community Traffic Safety Program, Maryland Legislators |
| Data Used | A variety of data available through the CODES project is used in the development of our standardized reports. The data include crash, EMS, emergency department , hospital, citation, licensing, toxicology, and vehicle registrations which are used individually and through the use of linkage to develop the most appropriate reports. Normally, the most recent three years of data are included in the analysis. Imputed data sets are used as appropriate and, with the development of the data model tools, will become increasingly incorporated into our annual reporting. |
| Methodology and Analytical Results | The methodology for the use of imputed data follows the model developed by the CODES work group. Other reports use simple frequencies, means, and medians. Comparisons are most commonly made using chi-square and t-tests. |
| Dissemination Formats | Data and reports are provided on paper and in electronic formats. These reports are also made available through the Internet at http://nsc.umaryland.edu. In addition, approximately 20 data presentations are given to various highway safety groups throughout the year in Maryland. |
| Impact, Follow-Up, or Later Development on the Targeted Issue | Through the comprehensive approach, CODES has played an increasingly crucial role in the development of the State's Strategic Highway Safety Plan and in the identification of emerging highway safety problems and the evaluation of current programs. |

## Massachusetts CODES
**UMassSAFE, University of Massachusetts**

www.ecs.umass.edu/umasssafe/codes.htm

State Application Presented at CODES Annual Meeting, 2006

Title: *Evaluation of Frequency and Injury Outcomes of Lane Departure Crashes*

Abstract:

Abstract: Lane departure crashes account for approximately 19 percent of all crashes in Massachusetts but almost 46 percent of crashes involving fatal injuries. A lane departure crash occurs when a vehicle leaves the travel lane resulting in a collision. Lane departure crashes typically involve running off the road onto the right or left shoulder and hitting a fixed object, such as a tree, a pole or even a parked vehicle. However, lane departures could involve crossing into an opposite lane and colliding with a vehicle traveling in the opposite direction. Although the vast majority of lane departure crashes are collisions with fixed objects, it can be argued that lane departures resulting in crashes between two moving vehicles are potentially more severe and costly. Therefore, countermeasure implementation strategies should consider not only frequency but also crash costs and severity.

Crash data alone are insufficient to effectively evaluate crash injury and cost outcomes. Additional injury data can be gathered by complementing crash data with medical information. The CODES links data collected at the crash scene to hospital databases that contain specific injury data; this provides an enhanced understanding of crash injury outcomes by tracking crash victims through the health care system. Massachusetts CODES data were analyzed to examine injury outcomes of lane departure crashes considering hospital charges and length of stay. Results of the analysis would provide a basis for determining where to implement countermeasures based on crash cost and severity rather than frequency alone.

| | |
|---|---|
| Contact | Heather Rothenberg, UMassSAFE, 413-577-4304 |
| Population Targeted | Requested in conjunction with other related analyses by the Massachusetts Highway Department; other targeted audiences include Safety Analysts and Traffic Engineers working on Mass. issues |
| Traffic Safety Issue Targeted | This analysis served to provide a preliminary overview of outcomes associated with lane departure crashes to guide programming and countermeasure implementation. Specifically, the differences in outcomes (charges and length of stay) for head-on crashes in comparison to single vehicle run-off-road crashes were examined. Although single vehicle run-off-road crashes are far more frequent, the hypothesis was that head-on collisions had higher charges and were more injurious; therefore, countermeasure implementation should consider not only crash frequency but also severity. |
| Requesting Office or Targeted Audience | Massachusetts Highway Department, Safety Analysts and Traffic Engineers |
| Description of linked data | Crash data were linked to inpatient discharge data for FY2003 (data available at time of linkage). Both high probability and multiply imputed links were used in the analysis. For the 12 months of linked data, the average number of links in the imputed data was 3,362 linked record pairs. Of those, the average number of pairs identified as cases resulting from lane departure crashes was 1,021 cases, which corresponds to 30 percent of all linked pairs. Furthermore, 240 cases were categorized as lane departure head on crashes (23 percent of lane departures) and 781 cases were categorized as single vehicle run-off-road crashes (77 percent of lane departures). |
| Analytical Results | Hospital charges were found to be higher for lane departure crashes than for all crash types ($18,460 versus $16,302). In addition, when examining the two specific categories, lane departure head on crashes showed higher hospital charges than single vehicle run-off-road crashes ($20,413 versus $17,918). In terms of length of stay in hospital associated with crash injuries, there were no differences between all crashes and lane departures crashes. However, median length of stay associated with injuries from lane departure head-on crashes was found to be one day longer than length of stay associated with single-vehicle run-off-road crashes. |

# Massachusetts CODES

**UMassSAFE, University of Massachusetts**

www.ecs.umass.edu/umasssafe/codes.htm

State Application Presented at CODES Annual Meeting, 2006

## Title: *Evaluation of Frequency and Injury Outcomes of Lane Departure Crashes*

| Analytical Results (continued) | Median charges and length of stay were also examined for injuries associated with lane departure crashes that were single vehicle run-off-road crashes broken down by the object the vehicle collided with. Hospital cases resulting from collisions with trees and run-off-road left or crossing median were the two types that had the highest median charges and also the highest median length of stay in hospital. |
|---|---|

| Object Collided With | Median Hospital Charges | Median Length of Stay (days) |
|---|---|---|
| Tree | $22,197 | 4.0 |
| Post, utility pole or light pole | $15,709 | 3.0 |
| Guardrail, median barrier or crash cushion | $16,289 | 3.0 |
| Curb, ditch, or embankment | $16,361 | 3.0 |
| Parked vehicle | $18,480 | 4.0 |
| Ran off road left or cross median | $21,281 | 3.0 |
| Ran off road right | $14,838 | 3.0 |
| Other/unknown | $17,827 | 3.5 |

| Dissemination Formats | Lane departure analysis results were presented primarily through a paper and are part of an ongoing dialogue between the University of Massachusetts and MassHighway regarding lane departure crashes. In addition, the results will be presented at the Annual Meeting of the Institute of Transportation Engineers in August 2006 and the paper will be included in the conferences Compendium of Papers. |
|---|---|
| Impact, Follow-Up, or Later Development on the Targeted Issue | Lane departure data analysis has been an ongoing project for MassHighway and the dialogue between the MassHighway safety analysts and UMassSAFE have included a discussion of how the CODES data can be used and the benefits of using CODES data in addition to traditional crash data. |
| Comments | Data from NHTSA's National Automotive Sampling System – Crashworthiness Data System (NASS-CDS) and FARS were used to evaluate lane departure crashes for 2002–2004 and to consider the two categories of lane departures defined for study.<br><br>Conclusions: Lane departure crashes result in a much higher proportion of severe injuries than other types of crashes, which makes them a serious concern in traffic safety. Countermeasures to reduce the frequency and severity of these types of collision should be a priority for transportation engineers. By complementing the analysis crash data with CODES data, additional information is given that allows for not only injury severity, but also the associated cost of injuries resulting from lane departure crashes. The two categories of lane departure crashes identified in this study were compared in terms of injury severity, hospital charges, and length of stay. Lane departure head-on crashes were found to result in more severe injuries and had higher associated costs and lengths of stay in hospital than single vehicle run-off-road crashes. Crash data showed higher frequency of single vehicle crashes colliding with a tree or a pole. CODES data showed that single vehicle collisions with trees and single vehicle running-off-left or crossing median had the highest hospital charges and days in hospital among single vehicle crashes. |

## Massachusetts CODES
### UMassSAFE, University of Massachusetts
www.ecs.umass.edu/umasssafe/codes.htm

State Application Presented at CODES Annual Meeting, 2007

Title: *Crash Outcomes for Older Occupants of Motor Vehicles in Crashes*

Abstract: Greater fragility has been shown to contribute to high fatality rates among older road users. This project used linked crash and hospital data to examine charges, length of stay, injury body region, and nature of injury for two categories of older vehicle occupants (65 to 84 and 85+) with adults (25 to 64) as comparison. Results were presented in several formats including fact sheets, report, and presentation to the Massachusetts CODES Advisory Board and Massachusetts Traffic Records Coordinating Committee. The findings of these analyses have provided information for the ongoing discussion around older vehicle occupants involved in crashes, specifically around older drivers. Previously, similar analyses were coupled with a review of policy and legislation around teen drivers to support decision-making around Massachusetts Graduated Driver Licensing laws. This review will integrate these results as part of the work aimed at identifying options for improved licensing practices for older drivers and assessing alternative transportation methods for older residents of Massachusetts (such as improved transit, etc).

| | |
|---|---|
| Contact | Heather Rothenberg, Project Coordinator/Lead Analyst, University of Massachusetts Traffic Safety Research Program, 413-577-4304, hrothenb@acad.umass.edu |
| Population/ Problem Targeted | Older drivers and older occupants of motor vehicles in crashes |
| Targeted Issue | Initial review of older driver/older person safety in Massachusetts with focus on policy issues around older drivers. |
| Requesting Office or Targeted Audience | Requested in conjunction with other related analyses by the Massachusetts Traffic Records Coordinating Committee, particularly the Registry of Motor Vehicles and Department of Public Health |
| Data Used | Data used: 2002-2003 for inpatient discharge, emergency department, crash, and citation. The use of multiple datasets allowed for an examination of older driver and older occupants of motor vehicles in a manner that would not have been possible using any one dataset alone. The use of linked data allowed for the consideration of older driver behavior as well as the crash outcomes associated with older people involved in crashes. The high quality associated with both hospital and citation data led to results that were considered highly reliable by members of the TRCC. |
| Methodology and Analytical Results | The analysis of data was presented using the results of the CODES linkage based on multiple imputations of missing links. Analyses included descriptive statistics for fields such as violations and crash frequency. In addition, the analysis of linked data considered charges for emergency department and inpatient separately due to scale (median charges) and the use of the Barell Matrix and odds ratios for six injury types and six injury locations on the body. Analyses considered older people (65 to 84) and much older (85+). When there were not enough cases for the much older category, the two were combined (65+). |
| | Some of the key findings are below; additional findings can be found in the report or fact sheet. |
| | TBI type 1 injuries were two times more likely in *older* occupants and four times more likely in *much older* occupants, when compared to *adults*. *Older* and *much older* vehicle occupants have a greater likelihood of suffering injuries to all body regions than *adults*, with the exception of spine and back injuries. |

**UMassSAFE, University of Massachusetts**

www.ecs.umass.edu/umasssafe/codes.htm

State Application Presented at CODES Annual Meeting, 2007

Title: *Crash Outcomes for Older Occupants of Motor Vehicles in Crashes*

| Methodology and Analytical Results (continued) | <ul><li>*Older* and *much older* vehicle occupants have lower inpatient charges resulting from crash injuries but have significantly higher emergency department charges than *adults* used as a comparison.</li><li>When considering charges, the only age group where there was a significant difference between drivers and passengers was for inpatient charges for *older* occupants. There are no notable differences in emergency department or inpatient charges between drivers and passengers for *adults* or *much older* occupants.</li><li>*Older* and *much older* occupants were also more likely to suffer fractures, internal organ injuries, open wounds and superficial injuries/contusions than *adults*; they were less likely to suffer sprains and strains.</li><li>The proportion of movement related violation (lane violations, failure to stop, turning violations) were notably higher for older drivers (65+) than for other drivers (<65).</li><li>Broad scope violations (leaving the scene, general violations, severe violations) accounted for a small percentage of violations for older drivers (65+) and other drivers (<65); the proportion of these violations was somewhat higher for older drivers (65+).</li><li>For both older drivers and other drivers, risky behavior violations (alcohol, seat belt use, and speeding) accounted for the largest proportion of violations issued. The proportion of risky behavior violations were similar for older drivers (65+) and other drivers (<65).</li></ul> |
|---|---|
| Dissemination Formats | Results of the analysis were presented in three formats:<br><br>Report: A report on the findings, specifically around the analysis of linked crash and hospital data, included a description of the linkage process, statistical methods, used, etc. as well as results. This document was designed for someone such as an analyst at another State agency. The use of specific information and more detailed formulas made this document most effective for the designated audience.<br><br>Fact Sheet: The fact sheet presented a brief summary of findings with no explanation of the linkage. This fact sheet was designed for someone in a State agency who has some experience "reading numbers" but who is not necessarily an analyst. The inclusion of "dollars" made this document especially appealing to those considering traffic safety from a policy perspective.<br><br>TRCC Presentation: A presentation was given to the TRCC outlining the process and the findings. This presentation included some basic information on the linkage and analysis but focused primarily on findings followed by a discussion on how TRCC members might use the information and data that might be added to the linkage to provide additional detail for future analyses. The combination of results and process made this most effective for the TRCC as TRCC members are interested in where the data came from, how it was analyzed, and what the results show. This presentation was the same as the presentation given to the CODES advisory board. |
| Impact, Follow-Up, or Later Development on the Targeted Issue | Based on the continued interest in older drivers and data-driven policy making, the Massachusetts Registry of Motor Vehicles has submitted a project problem statement to the Executive Office of Transportation's Cooperative Research Program to undertake a data-driven policy and legislative review. This review will integrate the findings of the data to identify options for improved licensing practices for older drivers and assess alternative transportation methods for older residents of Massachusetts. |

## Minnesota CODES

**Minnesota Department of Public Safety/Department of Health**

www.health.state.mn.us/injury/topic/topic.cfm?gcTopic=9

State Application Presented at CODES Annual Meeting, 2006

Title: *Estimating Minnesota Hospital Charge Savings With the Adoption of a Standard Enforcement Seat Belt Law*

| | |
|---|---|
| **Study Goals:** Measure hospital medical care charges associated with unrestrained motor vehicle occupants in Minnesota. Determine cost impact on government payer sources and estimate the likely impact of a standard (primary) enforcement seat belt law. | |
| Contact Person and Number | Anna Gaichas, 651-201-5478, anna.gaichas@state.mn.us <br> Mark Kinde, 651-201-5447, mark.kinde@state.mn.us <br> Scott Hedger, 651-201-7066, scott.hedger@state.mn.us <br> (Historical) Tina Folch, Minnesota Department of Public Safety Office of Traffic Safety |
| Population Targeted | The study focused on motor vehicle occupants and seat belt use. Emphasis was placed on two types of injuries caused by traffic crashes, traumatic brain injury and spinal cord injury. |
| Traffic Safety Issue Targeted | Unrestrained motor vehicle occupants in Minnesota. |
| Requesting Office or Targeted Audience | After numerous failed attempts to pass the primary seat belt enforcement law, it became apparent that a measure of the economic impact of not wearing a seat belt was necessary. The Minnesota Departments of Public Safety and Health along with the Minnesota Safety Council initiated this study. The target audience was policy makers, such as the Minnesota Governor and the Minnesota State Legislature. |
| Description of linked data | Individuals in Minnesota crashes that occurred during 2002 were linked with hospital emergency room and inpatient treatment information, the Traumatic Brain Injury Registry, and death certificate data. The linked data are referred to in this report as the 2002 CODES dataset. Although the TBI/SCI Registry and the hospital discharge dataset which were linked to the 2002 crash record have external cause of injury codes (E-codes) – for the purposes of the analysis, only cases that linked to a crash report were used. For this report, analysis only included people who were motor vehicle occupants. Seat belt use was imputed for those cases where seat belt use was unknown. Lastly, cost savings were only calculated for those cases where the individual was not belted and the linked hospital record had an injury coded (versus a non-injury coded). |
| Analytical Results | If Minnesota upgraded its seat belt law to standard enforcement in 2006 and reached a use rate of 94 percent, the cumulative charge savings to all government payer sources is projected to be $85.2 million by 2015. Injuries avoided in the first year alone would save Medicaid $3.4 million. The cumulative savings to Medicaid would be $70.9 million by 2015. |
| Dissemination Formats | A 26-page report and accompanying fact sheet were produced for distribution. On behalf of the Minnesota Seat Belt Coalition, the Minnesota Safety Council held a press conference at the State Capitol focusing on the results of the CODES report and the need for a belt law-upgrade to pass. Senate leaders spoke on the proposed primary seat belt bill and its potential economic impact in Minnesota. Later that day, the Senate passed the primary belt bill—with a floor vote showing a 2 to 1 margin in favor of the bill. Because of the passage of the Senate bill, media around the State disseminated the story by television, radio, and print communications. The fact sheet and full report were distributed through statewide e-mail distribution lists to law enforcement, public health educators, and other traffic safety advocates. The report was highlighted on the Office of Traffic Safety and Department of Health's Web sites and is currently available at http://www.dps.state.mn.us/ots/crashdata/CODES/MNCODES_BELTS_REPORT_March13_2006.pdf Data will continue to be used in various publications and presentations that discuss the economic impact of crashes. |
| Later development on the targeted issue | The report was used to gather support in the Minnesota Senate for passage of a primary belt bill, which was successfully signed into law on May 21, 2009. However, prior to 2009, the House of Representatives continued to display strong opposition to a belt law upgrade. During negotiations of the Transportation Omnibus Bill conference committee, the economic impact data was one strategy used by the Senate conference committee. |

# Minnesota CODES

**Minnesota Department of Public Safety/ Department of Health**

www.health.state.mn.us/injury/topic/topic.cfm?gcTopic=9

State Application Presented at CODES Annual Meeting, 2007

Title: *The Epidemiology of Motor Vehicle Crashes Involving 16–17-Year-Old Drivers in Minnesota and Associated Hospital Charges*

Abstract: Determine the per mile rate of teenage driver motor vehicle crash involvement and the characteristics of these crashes in Minnesota. Measure medical care charges and the severity of injuries associated with motor vehicle crashes involving teenage drivers.

| | |
|---|---|
| Contact | Scott Hedger, 651-201-7066, scott.hedger@state mn.us <br> Anna Gaichas, 651-201-5478, anna.gaichas@state.mn.us <br> Mark Kinde, 651-201-5447, mark kinde@state.mn.us <br> (Historical) Tina Folch, 651-201-7063, tina folch@state mn.us |
| Population/ Problem Targeted | Crash involvement and injury rates for 16–17-year-old drivers statewide |
| Targeted Issue | Demonstrate problems posed by inexperienced drivers and heighten public awareness of high crash involvement rate of 16-17-year-old drivers. |
| Requesting Office or Targeted Audience | The Minnesota Departments of Public Safety and Health and the Minnesota Safety Council determined that a fact sheet on the epidemiology of teen driver crashes would be useful supporting material for the legislative initiative to strengthen Minnesota's GDL law. The target audiences were the general public (especially parents and teens), safety advocates, the Minnesota State Legislature, and the Minnesota Governor. |
| Data Used | Nearly 27,000 Minnesota 2002 crash records were probabilistically linked with hospital emergency room and inpatient treatment information, the Traumatic Brain Injury Registry, and death certificate data, using ten linkage imputations. Linked data were assessed for quality by assuring high agreement between like variables on the crash and hospital sides, ensuring the percent missing was not higher than expected for variables of interest, and comparing the linked data to the crash data. |
| Methodology and Analytical Results | Hospital charges were calculated for linked/imputed cases. Miles driven were calculated using a Minnesota weighted proportion of regional responses to the National Household Travel Survey 2001 and limited to miles driving a car, van, SUV, or pickup truck. Maximum Abbreviated Injury Scale (MAIS) score for grouping severity was calculated using ICDMAP-90 software. Non-injured hospital cases with ICD-9-CM diagnosis codes other than 800-999 were included in the linkage and given their own category when analyzing severity. These cases have diagnoses that conceivably could follow a crash, e.g., observation following motor vehicle crash, neck/back pain, and chemical abuse. <br><br> Total hospital charges for 16- and 17-year-old drivers and for all people injured in a crash involving a 16- and 17-year-old driver were calculated. Overall, fatal, nighttime (10 p.m. – 4:59 a m.), and two-or-more-passenger crash involvement rates per miles driven were calculated for 16- and 17-year-old drivers and a comparison group of adult drivers (25 to 59 years old), using crash data. <br><br> Even though teens drive less than do adults, teenage drivers are overrepresented in motor vehicle crashes. In 2002, drivers age 16 and 17 comprised 3.2 percent of the State's licensed drivers and accounted for 2.1 percent of all the miles driven. Yet this group accounted for 8.7 percent of drivers in all crashes, 8.8 percent of drivers in injury crashes, and 8.6 percent of drivers in fatal crashes. The per miles driven fatal crash involvement rate of 16- and 17-year-old drivers was 4.5 times that of adult drivers. The per miles driven crash involvement rate of 16- and 17-year-old drivers was 3 times that of adult drivers. The per miles driven nighttime injury crash involvement rate of 16- and 17-year-old drivers was nearly 4 times that of adult drivers. |

| Minnesota CODES | |
| --- | --- |
| | **Department of Public Safety/ Department of Health** |
| | www.health.state.mn.us/injury/topic/topic.cfm?gcTopic=9 |
| | State Application Presented at CODES Annual Meeting, 2007 |
| | Title: *The Epidemiology of Motor Vehicle Crashes Involving 16–17-Year-Old Drivers in Minnesota and Associated Hospital Charges* |
| Methodology and Analytical Results (continued) | Sixteen- and 17-year-old drivers with two or more passengers in their vehicles were twice as likely to be involved in a fatal crash and 1.5 times more likely to be involved in an injury crash than adult drivers with two or more passengers. Inpatient and emergency department charges for teenage drivers in motor vehicle crashes were nearly $5 million and nearly $11 million for all people in motor vehicle crashes involving teenage drivers. Average hospital charges are greater for others in crashes involving 16- and 17-year-old drivers than the teen drivers. |
| Dissemination Formats | Initially, a presentation highlighting low seat belt use rates, over involvement in nighttime crashes and the relationship between number of passengers and crash involvement and severity was delivered at the annual Minnesota Towards Zero Deaths Conference in November 2006. The presentation was made available for traffic safety advocates and Public Health Liaisons to use as an educational tool through the Minnesota Office of Traffic Safety Web site.<br><br>Because teen crashes receive particular attention between prom and high school graduations, it was decided that a one-page fact sheet be developed for use during this time period. In May 2007, the fact sheet, *The Epidemiology of Motor Vehicle Crashes Involving 16–17-Year-Old Drivers in Minnesota and Associated Hospital Charges*, was finalized after receiving feedback from CODES board members. The informational sheet focused on defining major issues associated with teen driver crashes. The Department of Public Safety Communications Office worked in conjunction with the Minnesota Hospital Association and Department of Health in releasing a news advisory to statewide media outlets. The piece was picked up by major TV networks (lead story on one of the Twin Cities area nightly newscasts) and print papers statewide. The materials were also distributed via email to law enforcement agencies, public health advocates, other traffic safety partners statewide, and Traffic Records Coordinating Committee members.<br><br>After soliciting input from key advocates involved in teen traffic safety initiatives, it was determined that a full report exploring the potential impact of upgrading Minnesota's graduated drivers licensing law be developed.<br><br>The fact sheet mentioned is available for viewing at www.dps.state mn.us/ots/crashdata/CODES/TeenDrivers.pdf. |
| Impact, Follow-Up, or Later Development on the Targeted Issue | • Traffic safety advocates will incorporate the CODES information into their educational outreach efforts with parents, teens, the media, and policy decision makers.<br>• The Minnesota Legislature adopted a bill to strengthen Minnesota's Graduated Driver's License Law.<br>• Parents are encouraged to more closely monitor teen drivers and restrict teen driving during the evening and with passengers. |

**Missouri Department of Health and Senior Services**
www.dhss.mo.gov/MICA/index.html (Select motor vehicle crashes)
State Application Presented at CODES Annual Meeting, 2006
Title: *Safety Device Use for Children Age 4–8*

Abstract: Provided, on request, CODES data relating to the effect of safety device use on emergency department and inpatient charges, ejection, hospitalization, traumatic brain injury, and death for children age 4 to 8.

| | |
|---|---|
| Contact Person and Number | Mark Van Tuinen, 573-751-6300 |
| Population Targeted | Children age 4 to 8 |
| Traffic Safety Issue Targeted | Modifying the child safety seat law to require that children age 4 to 8 be secured in passenger restraint systems. |
| Requesting Office or Targeted Audience | Legislators, Hospital Administrators, Child Passenger Safety Advocates |
| Description of linked data | For this study, high-probability links (ED and inpatient; mortality beginning in 2003) were used, though a few from .80 to .89 are included based on inspection of person and crash identifiers. Data years 2001, 2003 were included. Over 90 percent of hospital ED and inpatient records have person name and E-codes; 99 percent have date of birth, sex, admit and discharge dates; up to 23 diagnoses are collected. All acute care hospitals report. Regarding crash records, 20,000 of 365,000 (5%) have few identifiers. |
| Analytical Results | Tables were prepared that showed statistically significant increased rates of ejection, hospitalization, TBI and death for nonusers of seat belts / child safety seats for children age 4 to 8. There were 3090 children age 4 to 8 for whom seat belt usage was known. Children who did not use seat belts were more likely to be ejected from the vehicle and admitted to the hospital. These children were also more likely to incur traumatic brain injuries or fatal injuries than children using seat belts. In 2003, the average inpatient charge for nonusers of seat belts was $16,240 higher than the average for users, and the median inpatient charge was $5252 higher. In 2001 the median inpatient charge was higher for nonusers. The average inpatient charge was slightly higher for users of seat belts but not significant. The average ED charge was higher for nonusers in 2003. The average ED charge was higher for nonusers in 2001 as well, but the difference was not statistically significant.<br><br>Government pay sources (primarily Medicaid) accounted for 61 percent of nonusers of seat belts, compared to 40 percent for users. Nonusers had over a million ($1,000,772) in charges for the Government pay source. |
| Dissemination Formats | Tables and text to explain the tables. |
| Impact, Follow-Up, or Later Development on the Targeted Issue | Hospital staff are now using the data to train safety seat technicians, advise families on the need for safety seats, and for grant applications. The bill was eventually passed. |

| Missouri CODES | |
|---|---|
| **Missouri Department of Health and Senior Services** | |
| www.dhss.mo.gov/MICA/index.html (Select motor vehicle crashes) | |
| State Application Presented at CODES Annual Meeting, 2006 | |
| Title: *Update of Web Application* | |

Abstract: The Missouri CODES Web application allows users to develop tables of CODES data which relate belt/helmet use, alcohol involvement, driver gender and age, crash type, speed zone, and vehicle type to level of medical care/death, ejection, and traumatic brain injury. Five years of CODES data is now on the Web site

| | |
|---|---|
| Contact Person and Number | Mark Van Tuinen, 573-751-6300 |
| Population Targeted | Drivers of motor vehicles |
| Traffic Safety Issue Targeted | Speeding, helmet and belt use, alcohol use, age, and sex of driver as related to health outcomes of death, TBI, ejection, hospitalization. |
| Requesting Office or Targeted Audience | Highway safety professionals, policymakers, legislators, community leaders. |
| Description of linked data | Probabilistic linkage - emergency departments and inpatient; mortality beginning in 2003 - though a few from .80 to .89 are included based on inspection person and crash identifiers. Data years 1993, 1996, 1999, 2001, and 2003 are included. Over 90 percent of hospital ED and inpatient records have person name and E-codes; 99 percent have date of birth, sex, admit and discharge dates; up to 23 diagnoses are collected. All acute care hospitals report. Regarding crash records, 20,000 of 365,000 (5%) have few identifiers. |
| Analytical Results | The Web site allows one to construct tables relating crash variables to medical care variables, as in showing the percent of helmeted versus unhelmeted motorcycle drivers who die, are admitted, visit the emergency room or receive no hospital care. |
| Dissemination Formats | The Web site is available to all individuals with Web access. |
| Impact, Follow-Up, or Later Development on the Targeted Issue | During January to April 2006, an average of 40 department and 279 outside users per month visited the CODES Web site, for a 4-month total of 161 department and 953 external visitors. |

# Missouri CODES

**Missouri Department of Health and Senior Services**

www.dhss.mo.gov/MICA/index.html (Select motor vehicle crashes)
State Application Presented at CODES Annual Meeting, 2007

Title: *CODES Data Highlights the High Cost of Not Buckling Up*

Abstract: CODES data were used to relate seat belt use to the level of care needed, ejection, urban/rural crash site, driver impairment, inpatient and emergency department (E/D) charges, and expected pay source.

| | |
|---|---|
| Contact Person and Number | Mark Van Tuinen, 573-751-6300 |
| Population/ Problem Targeted | Unbelted and belted drivers of passenger vehicles in crashes statewide |
| Targeted Issue | Statewide effort to get a primary seat belt law passed |
| Requesting Office or Targeted Audience | Legislator |
| Data Used | Probabilistic linkage for 2004 linked crash-inpatient-ED-mortality records. Crash data are from all crashes in Missouri in which an injury or $500 damage occurred. Inpatient-ED data are from all acute care hospitals in Missouri as well as from a number of hospitals in neighboring States. Mortality data are for all Missouri resident deaths regardless of where they died. Approximately 65 percent of inpatient and ED records noting an E-code in the 810-819 were linked to crash records. |
| Methodology and Analytical Results | Probabilistic linkages from an initial linkage were used. Inpatient-ED-mortality records were unduplicated prior to linkage, and likely repeat health care visits for the same person were omitted. Chi-square and t-tests were used to compare belted and unbelted drivers on the dependent variables. |
| Dissemination Formats | Tables with explanatory footnotes were emailed to the requestor, a member of the Traffic Records Coordinating Committee and one of the leaders in the statewide effort to generate support for a primary seat belt law. |
| Impact, Follow-Up, or Later Development on the Targeted Issue | Passage of the bill was unresolved at the time of this abstract. |

# Nebraska CODES

**Nebraska Department of Health and Human Services (NDHHS)**
**Division of Public Health**
http://www.dhhs.ne.gov/codes/
State Application Presented at CODES Annual Meeting, 2006

## Title: *Why Is It So Risky to Drive on the Roadways Where the Posted Speed Limit Is 50 Miles per Hour?*

Abstract: The crashes that occurred on roadways with a posted speed limit of 50 mph resulted in severe crash outcomes with the highest injury rate and a higher death rate than those crashes that took place on other roadways. This study examines the causes and consequences of crashes occurring on roads with posted speed limit 50 mph in Nebraska.

| | |
|---|---|
| Contact Person and Number | Ming Qu, 402-471-0566 |
| Population Targeted | Nebraska drivers |
| Traffic Safety Issue Targeted | Reduce the number of crashes that take place and prevent injuries from occurring to those involved in crashes. |
| Requesting Office or Targeted Audience | Analysis planned in conjunction with the State Highway Safety Office; other audiences includes State Department of Motor Vehicles and Driver Safety Schools |
| Description of linked data | CODES 1999–2000 results obtained by linking crash, EMS, and hospital discharge data using probabilistic linkage with acceptable linkage rates and limitations. |
| Analytical Results | Using descriptive analyses and multiple logistic regression modeling, the study indicated that, for those crashes that took place on roadways with a posted speed limit of 50 mph, there was a higher risk of death and severe injury. Contributing risk factors include involving teen drivers, low restraint use, alcohol, single-vehicle fixed objects, and/or lane departures resulting from speeding or driving too fast for conditions on two-lane roads with a gravel surface. |
| Dissemination Formats | • Made presentations at CODES Advisory Committee Meeting, the Statewide Highway Advocate Meeting, and the 2005 Traffic Records Forum. In addition, a special meeting was held at the Nebraska Office of Highway Safety to present this study and discuss recommendations for prevention<br>• Published articles about this study in two local newspapers.<br>• Published the paper in the *Journal of Safety Research*.<br>• Produced a fact sheet that was distributed at the Nebraska Department of Health and Human Services (NDHHS) Manager Meeting, and posted it on the NDHHS Web Site |
| Impact, Follow-Up, or Later Development on the Targeted Issue | Nebraska Office of Highway Safety took recommendation to add the gravel road driving training component into *Getting Your Driver's License in Nebraska: A Guide for Teens* |

# Nebraska CODES
## Nebraska Department of Health and Human Services (NDHHS)
## Division of Public Health
http://www.dhhs.ne.gov/codes/
State Application Presented at CODES Annual Meeting, 2007
## Title: *Comparing CrashesTthat Occurred in Nebraska Involving Nebraska and Non-Nebraska Drivers*

Abstract: Motor vehicle crashes occurring in Nebraska involved both State drivers and non-State drivers. Comparing the crashes occurring in Nebraska from 1999 to 2003 by the driver's State of residence, this study explores the patterns and the contributing risks factors of the crashes. The 1999–2003 data from the Nebraska CODES are used to describe the demographics of occupants involved in these crashes and crash characteristics. The study is designed to evaluate outcomes of these crashes and identify their causes—including drivers' and environmental contributing factors.

| | |
|---|---|
| Contact Person and Number | Ming Qu, 402-471-0566, ming.qu@nebraska.gov |
| Population/ Problem Targeted | Nebraska and non-Nebraska drivers who drive on the roadways in Nebraska |
| Targeted Issue | Reduce and prevent crashes occurring in Nebraska and their corresponding injuries and fatalities |
| Requesting Office or Targeted Audience | Analysis planned in conjunction with the State Highway Safety Office; other audiences include Traffic Records Coordinating Committee and Travel Information Center |
| Data Used | This study used the Nebraska 1999–2003 CODES data, a result of linking the crash to hospital discharge data with probabilistic linkage. |
| Methodology and Analytical Results | Using descriptive analysis, this study compared the Nebraska and non-Nebraska driver involved crashes and showed that crashes involving non-Nebraska drivers tended to be more severe, resulting in more deaths and serious injuries than crashes involving Nebraska drivers.  The crashes involving non-Nebraska drivers had the following characteristics:<ul><li>The most common reasons for the crashes were failure to yield, following too close and speeding, bad weather, and animals in roadways were cited as well.</li><li>The crashes were more likely to happen during the summer.</li><li>The crashes frequently occurred in rural areas or on State and interstate highways.</li><li>The vehicles driven by non-Nebraska drivers were:<ul><li>Most likely to be passenger cars</li><li>Disproportionately more likely to be heavy trucks or semi-trailer trucks.</li></ul></li></ul> |
| Dissemination Formats | <ul><li>Presentations were made at the CODES advisory committee meeting, the Statewide Highway Advocate meeting, and the 2006 Traffic Records Forum.</li><li>A special meeting was held at the Nebraska Office of Highway Safety to present this study and discuss recommendations for prevention.</li><li>Local newspapers published articles about this study.</li><li>Produced a fact sheet that was posted on the Nebraska Department of Health and Human Services' Web site. It is viewable online at http://www.dhhs.ne.gov/CODES/Non-NE-Drivers.pdf</li></ul> |
| Impact, Follow-Up, or Later Development on the Targeted Issue | Worked with the Nebraska Office of Highway Safety, Nebraska Injury Prevention Advisory Committee, and the media to send out the message to the Tourist Information Center, Truck Association, and State Patrol. |

| | New York CODES |
|---|---|
| | **New York State Department of Health, Bureau of Injury Prevention** |
| | State Application Presented at CODES Annual Meeting, 2006 |
| | Title: *Using CODES and Imputation to Demonstrate the Importance of Promoting Backseat Seat belt Use* |

Abstract: The study applied imputation methodology to address the seat belt use over-reporting and no-reporting problems in police crash reports and imputed the backseat seat belt use. The results were then examined in relation to various potential risk factors including driver's age, gender, and crash time. The belted and unbelted backseat passengers were compared in Barell injury groups, hospital charges, and length of hospital stay. The research was presented during the New York State Highway Safety Conference. New York State Association of Traffic Safety Boards (NYSATSB) members attended the meeting. The paper on backseat seat belt use was published in Journal of Safety Research.

| | |
|---|---|
| Contact Person and number | (Historical) Motao Zhu, former NYS CODES Project Manager<br>(Current) Michael Bauer, Susan Hardman, 518-473-1143 |
| Population Targeted | Backseat passengers 16 and older. |
| Traffic Safety Issue Targeted | Increase backseat seat belt use and promote backseat seat belt legislation. |
| Type of Decision Maker Targeted | New York State Association of Traffic Safety Boards |
| Requesting Office or Targeted Audience | The New York State, Bureau of Injury Prevention initiated this project to meet needs of the NYS traffic safety community. There has been a continual effort in the NYS traffic safety community to get backseat seat belt legislation passed. |
| Description of linked data | Linked Police Crash Reports and Hospital Discharge Data, 2002; imputation for backseat seat belt use; missing links were imputed. As limitations, sensitivity and specificity of imputation model are unknown. |
| Analytical Results | Statistical methodologies: descriptive analysis, multivariate analysis. The results:<br>• Unreliable reported seat belt use was set to missing and a logistic regression model was applied to impute backseat seat belt use. The imputation adjusted the backseat seat belt use from 69 percent to 35 percent.<br>• Backseat seat belt use was tabulated by various potential risk factors. Driver's alcohol involvement, being male, and nighttime were associated with decreased restraint use. Barell Injury Matrix, hospital charges and stay were tabulated by backseat seat belt use. Nonuse of backseat seat belts was related to higher rates in hospitalization, traumatic brain injury, and vertebral cord injury. |
| Dissemination Formats | A presentation was made during the New York State Highway Safety Conference with NYSATSB in attendance. It is a statewide organization for county traffic safety boards and agencies, companies, and individuals working in traffic safety and injury prevention. It proposes traffic safety legislation and offers traffic safety education programs. |
| Impact, Follow-Up, or Later Development on the Targeted Issue | NYSATSB are planning to conduct health educational campaigns to increase backseat seat belt use and promote a traffic safety law requiring passengers 16 or older to buckle up in the backseat. |
| Comments | Seat belt use recorded in police crash reports has been criticized for misclassification because occupants might falsely report the seat belt use to avoid traffic violation fines or tickets. New York CODES imputed missing values of seat belt use and adjusted the totals to account for over-reporting. Imputation adjusted 2002 New York backseat belt use from the recorded 69 percent to 35 percent, which was more consistent with estimated national rate of 47 percent. Driver's alcohol involvement, male gender, and nighttime crashes were associated with nonuse of seat belts by adult backseat occupants. Backseat belt use percentage after imputations was 8.1 percent for driver's alcohol involvement, 35.3 percent for no driver's alcohol involvement, 29.7 percent for male, 40 percent for female, 27.8 percent at night, and 38.2 percent during the day, respectively. |

# New York CODES

**New York State Department of Health, Bureau of Injury Prevention**

State Application Presented at CODES Annual Meeting, 2007

## Title: *Using Multiply Imputed CODES Data to Identify Risk Factors and Reveal Societal Costs in Teen Driving*

Abstract: The linked and multiply imputed 2005 Police Crash Reports, Emergency Department and Hospital Discharge Data were examined to determine the risk factors and societal costs for drivers who were aged 16–20. Teen drivers and 25–49-year-old drivers were compared for traffic crash and injury rates, emergency department visit rates, hospitalization rates, and crash contributing factors. These data were used to estimate the societal costs for vehicle occupants riding with teen drivers involved in crashes, and vehicle occupants involved in crashes when the teen driver was at fault or speeding. The analysis results were used to prioritize the contents for the New York State Department of Health, Bureau of Injury Prevention's annual training workshops for local health departments and traffic safety agencies. The CODES application on teen drivers was also presented during the training.

| | |
|---|---|
| Contact Person and Number | (Historical) Motao Zhu, former New York CODES Project Manager<br>(Current) Michael Bauer, Susan Hardman, 518-473-1143 |
| Population/ Problem Targeted | Teen drivers (16–20 years old) |
| Targeted Issue | Reduce traffic crashes, injuries, and societal costs due to teen drivers. |
| Requesting Office or Targeted Audience | New York's Bureau of Injury Prevention initiated this project to better understand the issues surrounding teen drivers. This is an important issue in New York and the information gathered from the CODES analysis will help the New York traffic safety community prioritize there agenda around this problem |
| Data Used | Data sources: crash, emergency department, inpatient<br>Data year: 2005<br>Quality, comprehensiveness, representativeness: missing links were imputed, missing values were imputed, multiply imputed data were analyzed |
| Methodology and Analytical Results | Statistical methodologies and descriptive analysis. The results:<br>• The crash-related emergency department visit and hospitalization per 1,000 licensed drivers were 15.7 and 1.2 for 16- to 20-year-old drivers, which were double the rates for 25- to 49-year-old drivers.<br>• The societal costs for vehicle occupants riding with teen drivers involved in crashes was $1.8 billion in 2005. |
| Dissemination Formats | 1) Tables and talks within the NYSDOH and the development of a training plan.<br>2) Oral presentation during the training focused on topics related to revealing the risk factors and societal costs for teen drivers for local health departments and traffic safety agencies. |
| Impact, Follow-Up, or Later Development on the Targeted Issue | 1) The NYSDOH and BIP used the CODES application to prioritize the training contents.<br>2) Local health departments and traffic safety agencies will use the CODES application to develop strategies to reduce teen driver crashes, injuries, and societal costs. |

# Ohio CODES

**Center for Injury Research and Policy, Nationwide Children's Hospital**

http://sharedoc.nchri.org/CIRP/Pages/CODES.aspx

State Application Presented at CODES Annual Meeting, 2006

## Title: *Boost Advisory Board Interest in CODES*

Abstract: Ohio recently completed its first CODES data linkage, and thus, new information regarding motor vehicle crashes and their medical and financial outcomes are now available. An advisory board meeting was held to announce this achievement. A presentation was given to review information about CODES and to show preliminary results to generate interest and to facilitate discussion. Discussions allowed the CODES team to hear the suggestions of traffic safety partners. The applications below were still in progress at the time of this abstract, as data linkage had only recently been completed.

| | |
|---|---|
| Contact Person and Number | Gary Smith, M.D., Dr.P.H., Principal Investigator, 614-722-2400 (Historical) Kristen Conner, M.P.H., former CODES Data Manager |
| Population Targeted | Restrained/Unrestrained, Alcohol Impaired, Teen Drivers |
| Traffic Safety Issue Targeted | Restrained/Unrestrained: Support for primary belt use. Alcohol Impaired: Support BAC .08 law. Teen Drivers: Support GDL law. |
| Requesting Office or Targeted Audience | Advisory Board (representation of groups with traffic safety interests in Ohio). Eventually will target legislators |
| Description of linked data | Preliminary results analysis used crash records linked with hospital inpatient data for 2002. Data were linked using multiple imputations of missing links. |
| Analytical Results | Linked imputed CODES data were analyzed using SAS. Median length of stay and median total hospital charges were reported. 1. Data Model for Restraint Use: "Restrained" included any indicator of belt use or a child safety seat. "Unrestrained" indicated that the vehicle occupant had no form of restraint use. These two categories were compared by age groupings. Results: Graphs of hospital charges were effective in displaying the benefits of restraint use. 2. Data Model for Alcohol Use: "Alcohol involved" included any person with an indication that alcohol or alcohol and drugs were suspected in a crash. "No alcohol involved" included people who had no alcohol exposure, who had been drinking but were not impaired, or where drugs only were suspected. Again, these two categories were compared by age groupings. Results: Length of stay showed no effects. Alcohol use had higher charges for most age groups. Data is reported for all occupants involved in a crash; however, the alcohol flag indicates alcohol use by drivers in crashes. 3. Data Model for Teen Drivers: Flagged for "Drivers" only, age 15 to 19. Compared single-occupant crashes (driver alone) to crashes with four total occupants. Results: Length of stay and charges showed no effects for number of occupants. This may be due to the small number of crashes with four occupants. Crashes with two or more occupants are compared with single-occupant crashes. |
| Dissemination Formats | A presentation was given that reviewed the origins of CODES, Ohio CODES goals, and linkage techniques. Results were presented in frequency tables and bar charts. Bar charts displaying differences between groups for hospital charges effectively communicated the results of analyses. |
| Impact, Follow-Up, or Later Development on the Targeted Issue | The CODES team noted the input from the advisory board. The CODES team is working closely with the Ohio Department of Public Safety to focus on analysis of topics of interest to the traffic safety community. New results will be presented when available. More useful results will be obtained when the above suggestions are taken into account during future analyses to generate CODES reports, fact sheets, etc. |

| Ohio CODES |
|---|
| **Center for Injury Research and Policy, Nationwide Children's Hospital** |
| http://sharedoc.nchri.org/CIRP/Pages/CODES.aspx |
| State Application Presented at CODES Annual Meeting, 2006 |
| Title: *The Impact of a Standard Enforcement Safety Belt Law on Fatalities and Hospital Charges in Ohio: An Analysis Using 2003 Ohio CODES Data* |

| | |
|---|---|
| Abstract: The effect that enactment of a standard enforcement seat belt law in Ohio would have on hospital charges and direct medical costs due to motor vehicle crashes was analyzed with a focus on the impact to the State's Medicaid system. | |

| | |
|---|---|
| Contact Person and Number | Gary Smith, M.D., Dr.P.H., Principal Investigator, 614-722-2400 (Historical) Kristen Conner, M.P.H., former CODES Data Manager |
| Population/ Problem Targeted | Unrestrained motor vehicle occupants in Ohio |
| Targeted Issue | Support for primary enforcement seat belt legislation |
| Requesting Office or Targeted Audience | Analysis was planned in conjunction with the Governor's Highway Safety Office to support Ohio Department of Public Safety and Ohio Seatbelt Coalition; handouts provided to Ohio legislators |
| Data Used | This analysis used crash records linked with hospital emergency department and inpatient data for 2003. Multiple imputation of missing links (five imputations) and multiple imputation for missing data were used to analyze the data. |
| Methodology and Analytical Results | Unrestrained motor vehicle crash victims in the linked database were divided into three categories based on the nature of injury: TBI, SCI, and all other injuries. In order to accurately apply NHTSA-reported seat belt effectiveness rates, the analysis was limited to occupants of a passenger car or light truck who were older than 4 with a motorist safety equipment code of lap belt only, lap/shoulder belt, child safety seat, or unknown. The resulting data set included 51,960 linked crash and hospital records with charges totaling $108.8 million (n=51,313). |
| | Results: Overall, 12 percent (n= 6,113) of our linked population was unbelted. Total hospital charges for the unbelted occupants was $29.2 million (n=6,031). Medicaid was the primary payer for 10 percent (n=5,612) of occupants. Using our estimated increase in seat belt use to 92 percent, the projected total direct medical cost savings to Medicaid over 10 years for hospitalizations that occur in 2007 alone was $15.4 million. Assuming passage of a standard enforcement seat belt law in 2007, cumulative costs savings to Ohio's Medicaid system over 10 years was projected to be more than $91.2 million. In addition, it was estimated that a minimum of 18 fatalities would be prevented in 1 year. Our estimates are considered underestimates as we were unable to ascertain costs associated with injuries treated in a non-hospital setting; only motor vehicle occupants older than 4 were included; and the costs associated with those motor vehicle crash victims who died at the scene were not included. |
| Dissemination Formats | A formal report and a brief, one-page summary sheet were prepared and submitted to the Ohio Department of Public Safety. The Governor's Highway Safety Office joined us in publicizing and sharing our results with the Governor and other legislators. The report and summary are posted at: http://sharedoc nchri.org/CIRP/CODES/Abstract_StdEnforcement.pdf http://sharedoc nchri.org/CIRP/CODES/Full%20Report_StdEnforcement.pdf . The Ohio Seatbelt Coalition used our findings on cost savings in a FAQ sheet developed for the most recent attempt to pass a primary enforcement law. That FAQ is at http://www.morpc.org/pdf/FAQs_media.pdf |
| Impact, Follow-Up, or Later Development on the Targeted Issue | The GHSO took note of the results and potential impact. The report was shared with the director of ODPS, the Governor of Ohio, and other legislators. The standard enforcement legislation was re-introduced in the Ohio legislature in 2009 but was defeated. |

# Rhode Island CODES

**Rhode Island Department of Health, Center for Health Data and Analysis**

www.health.ri.gov/chic/statistics/codes.php

State Application Presented at CODES Annual Meeting, 2006

Title: *Rhode Island CODES Provides Expert Participation*

Abstract: Rhode Island's CODES-linked data results and other highway safety and public health information has been used in support of stricter laws against driving under the influence.

| | |
|---|---|
| Contact Person and Number | Ted Donnelly, 401-222-5142, Edward.Donnelly@health.ri.gov |
| Population Targeted | All drivers in Rhode Island. |
| Traffic Safety Issue Targeted | Reduce the percentage of highway traffic fatalities that are alcohol-related. Breathalyzer refusals are high in Rhode Island due to weak law and light penalty. |
| Requesting Office or Targeted Audience | Legislators and policy-makers |
| Description of linked data | Rhode Island CODES links Crash, EMS, hospital, and fatality records data. Crash records are available for all people in reported crashes and hospital data is complete and of high quality. Rhode Island CODES has five years of linked records, from 2000 to 2004. |
| Analytical Results | Standard descriptions with median and mean hospital charges, length of stay, and mortality are reported as outcomes. Mortality reports, enhanced with data from the Fatality Analysis Reporting System, are presented. |
| Dissemination Formats | This application derived from a presentation at the CODES annual national Technical Assistance meeting of 2005.<br><br>Formal presentations were delivered to highlight the significant number of fatal crashes that were alcohol-related and call for stricter laws and more rigorous enforcement. Part of the argument was that although other New England States had improved on this measure in recent years, Rhode Island had not shown measurable improvement.<br><br>Substance Abuse and Mental Health Services Administration tables were also created showing that Rhode Island is among the States with highest rates of binge drinking and illicit drug use. At the same time, Rhode Islanders were among those least likely to see binge drinking and illicit drug use as problems. |
| Impact, Follow-Up, or Later Development on the Targeted Issue | Legislation to stiffen penalties for breathalyzer refusal is making its way through committee with multiple sponsors and strong support. |

| Rhode Island CODES |
|---|
| **Rhode Island Department of Health, Center for Health Data and Analysis** |
| www.health.ri.gov/chic/statistics/codes.php |
| State Application Presented at CODES Annual Meeting, 2007 |
| Title: *Elder Occupants in Motor Vehicle Crashes: Forecasting Health Burden* |

Abstract: Statistically, Rhode Island's older occupants are more likely to be hospitalized or fatally injured after a motor vehicle crash. Understanding the differences and considering the age distribution of the population can help in planning policies on future highway safety interventions.

| | |
|---|---|
| Contact Person and Number | Ted Donnelly, 401-222-5142, Edward.Donnelly@health ri.gov |
| Population/ Problem Targeted | Compares older occupants with other age groups by hospital outcomes, including length of stay, charges, payer, and discharge status. |
| Targeted Issue | Highway safety interventions will be targeted and the age structure of the population in the future will be considered. |
| Requesting Office or Targeted Audience | Rhode Island CODES Professional Advisory Committee, which includes data owners, public interest groups (e.g., Mother Against Drunk Driving), injury control, and law enforcement. |
| Data Used | This analysis used CODES probabilistic linkages of 2003 records from Crash, EMS, and hospital discharge data. Missing values were multiply imputed as were missing record linkages. |
| Methodology and Analytical Results | Analysis included the distribution of occupants in crashes by sex, age group, and level of care. Occupants admitted to the hospital differed in charges and length of stay by age group and payer. Restraint use is not uniformly associated with lower charges or shorter hospital stays. |
| Dissemination Formats | This material was presented at a number of venues, culminating with a meeting of the Rhode Island CODES Professional Advisory Committee. The Rhode Island CODES presentations emphasized the use of Bayesian probabilistic linkage without providing a detailed explanation of the method. Those who attended these presentations expressed interest in graphic representations of the progression of large and small age cohorts through the lifespan. |
| Impact, Follow-Up, or Later Development on the Targeted Issue | This analysis and presentation further supports the continued emphasis on interventions with older people. |

| South Carolina CODES | |
|---|---|
| **South Carolina State Budget and Control Board, Office of Research and Statistics** | |
| www.ors2.state.sc.us (select SC CODES project) | |
| State Application Presented at CODES Annual Meeting, 2006 | |
| Title: *Providing Information to Support Decision Making* | |
| Abstract: South Carolina CODES data are used each year to support proposed changes to traffic related laws that are debated in the South Carolina Legislature. | |
| Contact Person and Number | Tracy Joyce Smith 803-898-9948 |
| Population Targeted | Targeted problems are those in which traffic safety legislation is being considered: motorcycle riders and driving under the influence. |
| Traffic Safety Issue Targeted | Legislative issues being considered in current session:<br>1. Driving-under-the-influence crashes<br>&bull; Motor vehicle crash outcome by primary payer, State total, South Carolina, 2003.<br>&bull; Hospital and emergency department charges at the State and county level.<br>&bull; Motor vehicle crash outcome by driver gender for crashes with at least one driver under the influence, State and county level.<br>2. Motorcycle crashes<br>&bull; Estimated economic cost of motorcycle crashes, injuries, and fatalities in South Carolina for the last five years.<br>&bull; Estimated economic cost of motorcycle crashes, injuries and fatalities at signalized intersections.<br>&bull; Estimated economic impact from red light running crashes. |
| Requesting Office or Targeted Audience | The requesting office for the majority of requests is the SC DOT, Office of Highway Safety. The OHS then combines our analysis with information gathered from other sources to create profiles at the State and county levels. The OHS then distributes these profiles to the appropriate traffic safety stakeholders, legislators, etc. Legislators receive information specific to their represented counties, etc. |
| Description of linked data | The South Carolina CODES project links crash reports with inpatient hospitalizations and emergency department visits from 1998 to 2000, which included emergency medical services as well. South Carolina's data linkage is of a high quality due to the identifiers available to the project. These links are representative of the population based on statistical testing of these data. |
| Analytical Results | The analytical reports for these various legislative initiatives were customized to the audience. Reports addressing the topic were targeted to legislator's districts, county safety coalitions, or other geographic areas of interest. The ability to customize specific safety reports on short notice makes the information "local." |
| Dissemination Formats | The Office of Research and Statistics provides the data to the South Carolina Department of Transportation and the Department of Public Safety, both partners in the CODES project. Resources from these organizations tailor the information to fact sheets, reports, PowerPoint presentations, and other means of dissemination. |
| Impact, Follow-Up, or Later Development on the Targeted Issue | Linked CODES data was instrumental in the passage of legislation to raise the fine for children who were not properly restrained. South Carolina Governor Mark Sanford vetoed legislation raising the fine from $25 to $150. Both the House and Senate over-rode the veto and the legislation has become law.<br><br>Reports on motorcycle issues were requested by the House of Representatives to address issues related to red lights, redefining classes of motor vehicles, and evaluating the safety issues for a variety of legislation. Reports on motorcycle use supported changes in legislation, which included revising definitions for types of motor vehicles, addressing the increase in injuries for people under 18, and other safety issues. |

## South Carolina CODES

**South Carolina State Budget and Control Board, Office of Research and Statistics**

www.ors2.state.sc.us (select SC CODES project)

State Application Presented at CODES Annual Meeting, 2007

### Title: *The Continuing Saga of DUI Legislation*

Abstract: CODES data is used to develop various media products that are distributed to members of the South Carolina legislature and traffic safety decision-makers to illustrate the medical and financial consequences of under-the-influence drivers involved in crashes. The data is analyzed at a local level to show local consequences.

| | |
|---|---|
| Contact Person and Number | Tracy Joyce Smith, 803-898-9948 |
| Population/ Problem Targeted | The State-specific application targets driving under the influence. |
| Targeted Issue | The South Carolina application provides data that support revising the current legislation (per South Carolina's DUI law) to clarify ambiguities in the original legislation. |
| Requesting Office or Targeted Audience | The requesting office for most requests such as this is the South Carolina Dept. of Transportation, Office of Highway Safety, which combines our analysis with information gathered from other sources to create profiles at the State and county levels and then distributes these profiles to the appropriate traffic safety stakeholders, such as the South Carolina legislators; other targeted audiences include groups for DUI legislation (e.g., Mothers Against Drunk Driving). |
| Data Used | South Carolina's State-specific application includes linked data for 2001 through 2004, which includes crash records, inpatient hospitalizations, and emergency department visits. The high-probability linked data was used for this application. Analysis done on both high-probability and imputed datasets showed that the identifiers used in the linkage are highly discriminatory, corresponding to the imputed data. |
| Methodology and Analytical Results | Reports were generated for each of the 46 counties in South Carolina. Each report contained:<br>• Number of impaired drivers involved; number and percent linked;<br>• Total and mean charges by level of care (emergency department versus inpatient);<br>• Total and mean length of stay (inpatient only);<br>• Primary pay source (Medicaid, Medicare, insurance, self/indigent).<br>The objective was to show that all counties, large and small, urban and rural, are affected by crashes where drivers are under the influence. |
| Dissemination Formats | The South Carolina Office of Research and Statistics is a partner agency with the South Carolina Department of Transportation (SC DOT). The reports primarily are in Excel spreadsheets and occasionally as a static SAS report; all information is electronically submitted to the SC DOT. The SC DOT, in turn, reformats and presents the information to legislators and advocacy groups for use in legislative hearings and other educational endeavors. The objective is to show what is happening in each legislator's specific county. |
| Impact, Follow-Up, or Later Development on the Targeted Issue | The debate on stiffer legislation for those driving under the influence is an ongoing effort, much like the primary seat belt enforcement that took over five years to pass. A bill designed to close many of the loopholes and enact stiffer penalties in the existing DUI laws, which was introduced during last year's session, was not passed. There will be another attempt to secure passage of this bill next session.<br><br>The legislature did pass, and the Governor did sign, a comprehensive bill to reduce underage drinking and underage access to alcohol. The bill also included strengthened provisions for the use of alcohol ignition interlock devices.<br><br>The CODES data (for drivers age 15 to 20 who were DUI) have been used extensively in court testimony during the penalty phase of trials for vendors who have had repeat sales of alcohol to minors. |

| | |
|---|---|
| **Utah CODES** | |
| **University of Utah Intermountain Injury Control Research Center** | |
| http://www.utcodes.org | |
| State Application Presented at CODES Annual Meeting, 2006 | |
| Title: *Using CODES Data to Strengthen Utah's GDL Laws* | |

Abstract: During the 1999 and 2000 Utah legislative sessions several aspects of a graduated driver licensing program were instituted. These included a mandatory minimum number of practice hours, as well as nighttime and passenger restrictions. Using a combination of crash and citation rates per licensed teen driver, Utah CODES studied the effectiveness of these new laws. The results found that while there was an overall reduction in the crash rates for 16-year-olds, this reduction was much smaller than what has been observed in other States. In addition, there was no reduction in the rate of crashes resulting in an emergency department visit, hospital admission, or death. Finally, it was found that there were very few citations issued for violating the GDL laws. The results were shared in presentations at national conferences and published in *Annals of Emergency Medicine*. These findings were then published in newspapers and aired on local news outlets. Utah traffic safety advocates asked for the findings to be presented in a fact sheet form to support legislation to strengthen the GDL system. During the 2006 legislative session, the fact sheet was distributed to all legislators and the findings were used in testimony to support the new GDL legislation. The GDL bill passed both the House of Representatives and the Senate and was signed into law by the Governor.

| | |
|---|---|
| Contact Person and Number | Larry Cook, 801-585-9760, larry.cook@hsc.utah.edu |
| Population Targeted | Licensed novice 16-year-old drivers. |
| Traffic Safety Issue Targeted | Utah's GDL system |
| Requesting Office or Targeted Audience | This work was planned in conjunction with the Coalition for Utah Traffic Safety (of which the State Highway Safety Office and Utah Department of Health are members), the Legislative Task Force for Traffic Safety, and the child advocate at Primary Children's Medical Center. Target audience also includes the Governor and the Legislature. |
| Description of linked data | We studied crashes involving teenage drivers between 1996 and 2001. These crashes were probabilistically linked to driver license, citation/conviction, emergency department, and hospital inpatient data. All data sources are statewide databases. Imputation was not used for this analysis. |
| Analytical Results | Outcomes examined included overall crash rates, nighttime crashes, crash severity indicators, seat belt usage, licensure status, and citations. Rate ratios, chi-square tests, and interventional time series analyses were used to assess changes before and after GDL implementation. The results showed that there were 27,304 16-year-old driver crashes during the study period. The overall crash rate per 1,000 licensed 16-year-old drivers decreased by 5 percent, and a time-series analysis showed a reduction of 0.8 crashes per month per 1,000 licensed drivers after GDL implementation. The nighttime crash rate did not change. In addition, there was no decrease in the rates for crashes resulting in emergency department visits, hospital admissions, or death. These results pale in comparison to some States that have reported reductions in crashes between 20 percent and 30 percent. Few GDL licensing citations were issued by law enforcement. In addition, only five citations were written for violation of GDL. |
| Dissemination Formats | Originally, the results were presented at the American Public Health Association's annual meeting. A manuscript was developed from this presentation, which was published in *Annals of Emergency Medicine*. The findings were also published in several newspapers and aired on local television news programs. As a result, local traffic safety advocates asked us to prepare some fact sheets about GDL along with talking points. The fact sheets were given to legislators and the data were presented in committee meetings. Since the data was published in a peer-reviewed journal, the results had more legitimacy for many of the legislators. |
| Impact, Follow-Up, or Later Development on the Targeted Issue | The new GDL bill passed both the Utah House and Senate and has been signed into law. Ongoing efforts will be made to analyze the impact of the new law in the coming years. |
| Comments | The results suggest that GDL may have contributed to a reduction in young driver crashes, but the effects were minimal compared with those shown in many other GDL evaluations. It appears that the GDL is not being enforced as expected. |

## Utah CODES

**University of Utah Intermountain Injury Control Research Center**

http://www.utcodes.org

State Application Presented at CODES Annual Meeting, 2007

### Title: *Primary Safety Belt Enforcement Efforts*

Abstract: Utah's current seat belt law only provides for secondary enforcement for adults. This year Utah CODES continued its ongoing efforts to support primary seat belt legislation. This support came in form of filling data requests, providing talking points, distributing fact sheets, and giving presentations. Working with an Emergency Medicine physician at Primary Children's Hospital, the researchers examined injuries to children riding with adults stratified by belt use status. The research found differences in injury patterns for children riding with unbuckled adults compared to children riding with buckled adults. In addition, when the analysis is limited to only restrained children, children with unbuckled adults are more likely to be hospitalized compared to children riding with buckled adults. These findings were submitted as abstracts titled, "Are adults who use child safety restraints for children using seat belts themselves?" and presented at the Traffic Records Forum meeting in Palm Desert, CA, and the American Public Health Association meeting in Boston, MA. Efforts are also underway to prepare this research for publication.

To prepare for the legislative session, Utah CODES provided talking points to safety advocates to use in their testimony and meetings with legislators. Also, Utah CODES prepared a fact sheet titled, "The Cost of Being Unbuckled," which was given to every legislator. The primary seat belt bill passed the Senate and for the first time made it to the floor of the House. However, the final vote did not pass in the House.

| | |
|---|---|
| Contact Person and Number | Larry Cook, 801-585-9760, larry.cook@hsc.utah.edu |
| Population/ Problem Targeted | Crash injury rates statewide |
| Targeted Issue | Advocate for primary seat belt use |
| Requesting Office or Targeted Audience | This analysis was conducted at the request of the Coalition for Utah Traffic Safety (of which the State Highway Safety Office and Utah Department of Health are members); Primary Children's Medical Center, and a physician testifying on behalf of the American Academy of Pediatrics. Target audience also includes legislators. |
| Data Used | Crash data linked to emergency department and inpatient from 1996 to 2004 were used. The data include all occupants involved in crashes and probabilistic linkage was used. |
| Methodology and Analytical Results | Utah CODES used a combination of logistic regression and descriptive statistics to analyze the importance of seat belts at reducing injuries. Utah CODES found that young adults were more likely to restrain the children in their vehicles but not themselves. Also, older adults were the most likely group to restrain themselves but not the children in their vehicles. Utah CODES also found that among restrained children, those riding in vehicles with unrestrained adults were more likely to be injured and link to a hospital record compared to restrained children riding with restrained adults. The Utah CODES fact sheet showed that nearly two-thirds of the charges billed to government insurance came from unrestrained occupants. Also, the average charges were 25 percent higher for unrestrained occupants compared to restrained occupants. Finally, unrestrained occupants were six times more likely to be admitted to a hospital following a crash and three times more likely to be treated at the ED compared to restrained occupants. |
| Dissemination Formats | The results comparing injury outcomes between restrained children stratified by adult seat belt use was presented at two national meetings: the Traffic Records Forum and the American Public Health Association. The fact sheets were also given to traffic safety advocates in Utah as well as given to individual legislators. |
| Impact, Follow-Up, or Later Development on the Targeted Issue | Information from both the presentation and fact sheets were presented during this year's legislative session. As has been the case for the past two years, the primary seat belt bill passed the Senate. However, the bill was defeated 33 to 39 in the House. |

# Virginia CODES

**Virginia Dept. of Motor Vehicles, VA Highway Safety Office,**
**Traffic Records Management, Reporting and Analysis**

www.vacodes.org

State Application Presented at CODES Annual Meeting, 2006

## Title: *Virginia CODES Web site www.vacodes.org*

Abstract: The Virginia CODES program was established to reduce deaths and injuries from motor vehicle crashes by providing relevant and actionable information from linked crash, EMS, and hospital discharge data. The primary means for accomplishing these goals is through informed transportation safety policy and public education.

The Virginia CODES program designed a freestanding Web site to communicate the results of linked crash information to the public. Web sites from across the country were studied to identify ways to present information in a variety of formats for different audiences. An independent Web site was created to highlight the importance of this information as an independent and objective information source for all interested parties.

Following review of other State CODES Web sites, active involvement by the CODES Board of Directors and CODES advisory board, www.vacodes.org was launched in May 2006.

| Contact Person and Number | Angelisa Jennings 804-367-2026 |
| --- | --- |
| Population Targeted | Information is developed to support report creation by geographic area, gender, age, vehicle type, major factors, outcomes, and other groupings. |
| Traffic Safety Issue Targeted | Major areas regarding crashes are currently available for review. There is also an online query tool to allow users to create a custom report to the user's specifications. |
| Requesting Office or Targeted Audience | Transportation safety professionals, legislators, the general public. |
| Description of Linked Data | The linked data included 2001, 2002, and 2003 Virginia DMV crash data, EMS data, and hospital discharge data. |
| Analytical Results | Many variables—such as injury rates, fatality rates, length of stay, hospitalization rates, and total hospital charges—are available for analysis by area of interest. |
| Dissemination Formats | Ready-to-use reports are instant-access reports covering a variety of crash areas from age to vehicle type. These reports are available as a PDF document.<br><br>Create-a-Report is a function for instant development of custom reports for specific questions by the Web visitor. These reports may currently be printed. Under development is a capability for e-mail and the creation of an Excel file or a PDF document.<br><br>Fact Sheets are being developed on major areas of interest such as driver distraction, teen driving, and speed as a factor in crashes and alcohol use.<br><br>Research Briefs are planned using topic and background information developed for CODES States. |
| Impact, resulting action, or Impact, Follow-Up, or Later Development on the Targeted Issue | Virginia's General Assembly will begin its next session in January. The impact of CODES data on informing legislation as well as policy will be monitored. Use of Virginia CODES data by advocacy organizations will also be tracked. |

# Virginia CODES

**Virginia Dept. of Motor Vehicles, VA Highway Safety Office,**
**Traffic Records Management, Reporting and Analysis**
www.vacodes.org
State Application Presented at CODES Annual Meeting, 2007
Title: *408 Funding Award for Virginia's Traffic Safety Information System*

Abstract: Virginia CODES was a significant factor in the Commonwealth of Virginia's 408 Strategic Plan for Traffic Safety Information System. This plan, which secured an $855,000 Federal funding award, listed three major projects that were key to enhancing the State's traffic records information systems over the next three years. CODES serves as a key database to provide Virginia with medical costs related to traffic crashes in Virginia.

| | |
|---|---|
| Contact Person and Number | Angelisa Jennings 804-367-2026 |
| Population Targeted | Crash injury rates statewide and by cities, counties and towns, vehicle safety, data quality, and a means to disseminate information to the public |
| Traffic Safety Issue Targeted | Reduce injuries and severity, improve traffic records, and inform the public of injury costs |
| Requesting Office or Targeted Audience | A study developed for the Governor of Virginia, members of the Virginia General Assembly, Traffic Records Coordinating Committee, State Highway Safety Office, Injury Control Office and the public |
| Data Used | The linked and imputed data included 2001 through 2005 Virginia DMV crash data, EMS data and hospital discharge data |
| Analytical Results | The Bayesian Method, which is a branch of logic that looks at solving problems by using probabilities, has many advantages over more traditional methods of direct one-to-one linkage and may be the only way to solve problems with known missing data elements. In CODES, three data sets are imported and standardized: crash (DMV), EMS, and hospital (VHI). These data sets are then matched together in the following way: 1. Crash (DMV) data to EMS data 2. EMS data to Hospital (VHI) data 3. Crash (DMV) data to Hospital (VHI) data All three of these data sets have the possibility of missing or incorrect data fields. CODES performs the Markov Chain Monte Carlo imputation with crash (DMV) data, EMS and hospital (VHI) data. By running 5 to 10 imputations, data sets can be produced with the same results as if they did not have any missing or incorrect information within the calculable error rate of standard deviation among all of the imputed data sets. |
| Dissemination Formats | • Virginia's CODES Web site, www.vacodes.org, was launched in May 2006 and currently contains five years of data from which the user can select standard reports or create online queries based on selected criteria • New to the Web site are fact sheets and posters that can be printed and displayed in emergency departments, schools, DMV locations, or other public offices to disseminate detailed, specific messages about general crash facts, motorcycles, alcohol, and speeding. |
| Impact, Follow-Up, or Later Development on the Targeted Issue | CODES data and information assisted DMV in securing the 408 Federal funding for their strategic plan for Traffic Safety Information System; www.vacodes.org is also a beneficial portal of information used by State police, Mothers Against Drunk Driving, and other groups who want to quickly obtain specific and statewide data. |

# III. APPENDICES

## Appendix 1: Frequently Asked Questions Regarding CODES

### 1. What Is the Crash Outcome Data Evaluation System?

CODES, a program facilitated by the U. S. Department of Transportation's National Highway Traffic Safety Administration, provides software and technical assistance to States to study the population of all occupants in police reported crashes and to use the results to improve traffic safety. Crashes not meeting the State's police reporting threshold or out-of-State crashes involving victims treated in-State are excluded. CODES evolved from a congressional direction and has become institutionalized in many States, producing data analyses on crash outcomes in terms of mortality, morbidity, injury severity, and health care costs. To date, NHTSA has funded 30 States to implement CODES.

Police-reported crash data is the major source of population-based information about crashes statewide. Thus, they are crucial for traffic safety decision-making. However, because the impact of the crash on the occupants of the vehicles involved is not usually known at the scene, crash data does not include the injury outcome information that traffic safety analyses could use to evaluate effectiveness in terms of decreased mortality, morbidity, injury severity, and hospital costs. As with all routinely collected State data used for administrative purposes, they also are limited by the reporting threshold and by missing and inaccurate data. Additional information is needed to identify such characteristics of the person, vehicle, and/or event that are likely to result in specific injury outcomes for comparative purposes or for population estimates.

In contrast, the injury data files are created for public health purposes. They include medical details about the type and severity of the injury and the subsequent costs (billed charges) for all people treated for an injury, regardless of the cause of that injury. Injury data describes the injury outcome at the location of treatment, either at the scene or en route, at the emergency department or after admission as an inpatient. Different entities manage and control access to this data. Injury data does not contain information on the crash that generated the injury. Thus, traffic safety professionals cannot use the injury data alone to obtain the outcome information it needs to target State resources.

CODES is designed to address integrated needs by linking person-specific motor vehicle crash and injury State data to obtain the crash injury outcome information needed to analyze where improvements could be made to traffic safety. In addition, the linkage techniques enable the inclusion of other traffic safety State data — such as vehicle registration, driver licensing and citation data — that expand the comprehensiveness of the crash outcome information generated.

### 2. What does CODES provide that other crash data sets cannot provide?

CODES linked crash outcome data are a unique resource because they identify crash characteristics for both the injured and the non-injured. Analyses are less likely to be biased when data include characteristics of people involved in crashes who have unexpected outcomes: people who are injured in spite of using safety equipment and people who are not injured in spite of not using safety equipment.

The CODES program is designed to enhance the existing State data without the expense of additional data collection. For example, the crash outcome linkage may provide EMS and hospitals with the time of the crash in order to calculate a measure of the responsiveness of the trauma system; roadway inventories may be enhanced with the inclusion of injury type and severity by location; and licensing data may be enhanced when driver information is linked to the injury severity and health care costs associated with factors such as driving under the influence, aggressive driving, or speeding.

In pursuit of traffic safety goals, the CODES program promotes collaboration between the State traffic safety and health communities. Owners of the State crash and injury data serve on a CODES board of directors, which is responsible for ensuring that State data is available for linkage and for developing the policies that control release of the linked data in compliance with State privacy legislation/regulations. The success of these proactive partnerships spills over into other areas of traffic safety, which also depend upon a collaborative approach to improve crash outcome. This collaborative approach is consistent with the NHTSA's Highway Safety Program Guidelines and the Data Improvement Grants. States with CODES programs have found that CODES helps meet guidelines for traffic records assessments and had already established much of the structure required for data improvement funding.

CODES data and analyses have also been useful to promote public policy, including safety legislation. Because the CODES crash-outcome data are State-specific, they are more relevant to State legislators assessing primary belt laws, graduated drivers license provisions, or helmet use legislation. The ability to compare State-specific results to national estimates provides further clarification about the need for action.

### 3. How does CODES generate the linked crash outcome data?

Each State links person-specific crash records to the statewide ambulance run reports , hospital emergency department and inpatient records, and death certificate records, all of which are also person-specific. Few States include in the State data unique identifiers, such as social security numbers. Instead, indirect identifiers that discriminate among the events and the people involved are matched. Some States augment the person-specific crash outcome data with driver-specific data from the State licensing files, vehicle-specific data from the State registration data files, and roadway-specific data from the roadway inventory data files to facilitate the linkages.

The linkage is a sophisticated process. In the real world, it cannot be known for certain which crash and injury records are true matches. The lack of unique identifiers, weak indirect identifiers, records (crash or injury) missing for occupants known to have been injured—in addition to the expected problems of missing, and/or inaccurate data—all contribute uncertainty. After evaluating the quality of the State data, State CODES programs implement advanced methods of linkage using CODES2000 software, which estimates the probability that a possible record pair is a valid match.

### 4. How does CODES handle missing links?

Not all valid matches have high probabilities. This occurs when either the crash or injury record is missing or when the identifiers are unable to discriminate among the crashes and the people involved or are weakened because of missing, inaccurate or inconsistent values. Conclusions based only on high probability linked pairs cannot be presented as representative of the population. Linked pairs, excluded because of low probabilities caused by weak identifiers or incomplete data, may in fact be valid. To compensate for the imperfect data, CODES constructs ("imputes") multiple sets of data that can be used to statistically summarize estimates about the crash population. These estimates are representative of the population from which they were derived, just as a scientifically selected survey sample is representative of the population from which it was drawn.

## 5. How are imputed datasets analyzed?

Once the missing links have been identified, standard techniques for handling missing values are used to analyze the linked datasets. In SAS, the procedures used are PROC MI and PROC MIANALYZE. These techniques provide confidence intervals that accurately reflect uncertainty caused by missing data.

## 6. Who can access the CODES data?

CODES data remains in the States due to various issues concerning privacy and data ownership. Access to the CODES statewide linked data is controlled by the data owners in the States through the State CODES programs. Many sites maintain Web sites where CODES data can be accessed in the form of data runs or aggregate reports. When CODES data are used for NHTSA-sponsored studies, the submissions are in standardized analysis variables defined and derived specifically for the study, rather than an entire data set. NHTSA does not maintain a national CODES database.

**Appendix 2: Probabilistic Linkage Using Multiple Imputation**

The Crash Outcome Data Evaluation System links crash reports to injury outcome records—such as ambulance run reports (EMS), emergency department, or hospital discharge records—in order to evaluate injuries and medical charges associated with crashes. In addition, other traffic safety data sets including roadway inventory, vehicle registration, driver licensing and citations, and insurance claims may also be linked to provide a more comprehensive picture. Most CODES data sets do not have common unique identifiers. Consequently, CODES applies a statistical methodology to link the data sets. The probability that two records are a true link is determined by comparing all event characteristics (e.g., date and place) and all person characteristics (e.g., age and sex) that are common to both records. These characteristics are called *quasi-identifiers*.

CODES record linkage is conducted using CODES2000, commercially available software that implements an extension of Fellegi and Sunter's statistical theory of record linkage (Fellegi & Sunter, 1969; McGlincy, 2004, 2006). CODES2000 determines the posterior odds for a true link by applying Bayes' rule for odds (Gelman et al., 2004, pg. 9), "the posterior odds are equal to the prior odds multiplied by the likelihood ratio." Parameters of the linkage model are determined using Markov Chain Monte Carlo data augmentation (Schafer, 1997, pg. 72). CODES linkage concepts are summarized in Table A-2-1.

| Table A-2-1. CODES Linkage Concepts (Pr X means Probability of X) | | |
|---|---|---|
| **Concept** | **Definition** | **Calculation** |
| Probabilistic Record Linkage | Bayes' Rule for Odds applied to record linkage: Posterior odds for a true match equal the prior odds multiplied by the likelihood ratio. | Posterior Odds = (M / U) X (m / u) |
| Prior Odds for a True Match | Odds for a true match estimated from prior information. Posterior odds after comparing one match field become prior odds for next. | M / U = Estimated # of Matched Pairs / Estimated # of Unmatched Pairs |
| m Probability | Conditional probability for a comparison result (agreement, disagreement, or missing) for true matched pairs. | m Agreement = Pr(Reported) X Pr(Correct ) X Pr (Field has Given Value for Matched Population) |
| u Probability | Conditional probability for a comparison result (agreement, disagreement, or missing) for true unmatched pairs. | u Agreement = Pr(Reported) X Pr(Correct) X Pr (Field has Given Value for Crash Population) X Pr (Field has Given Value for Hospital Population) |
| Likelihood Ratio for a True Match | Likelihood for comparison result for true matched pairs/Likelihood for comparison result for true unmatched pairs. | Likelihood Ratio Agreement = (m Agreement / Pr (Agreement)) / (u Agreement / Pr (Agreement)) = m Agreement / u Agreement |

Missing values and reporting errors in the data collection processes may lead to low probabilities being assigned to many true matches. If only high-probability links are selected, then low-probability false negatives can make selected links unrepresentative of the total population of true linked pairs. To be able to include these low-probability matches in outcome studies, CODES2000 completes five linkage imputations; that is, missing links are determined five times resulting in five complete datasets. (Note that multiple imputation does not attempt to identify each missing link but instead constructs samples representative of the distribution of low to high probability links. As a result, analyses yield valid statistical inferences that reflect the uncertainty associated with having low-probability true links.) Standard statistical analyses are performed on each of the five datasets and then combined to produce final results using procedures in SAS.

# References

Fellegi, I.P., & Sunter, A.B. (1969). A Theory for Record Linkage. *Journal of the American Statistical Association*, 64, 1183–1210.

McGlincy, M. A. (2004, August 9). Bayesian Record Linkage Methodology for Multiple Imputation for Missing Links. Joint Statistical Meeting, 2004.

McGlincy, M. A. (2006, August 10). Using Test Databases to Evaluate Record Linkage Models and Train Linkage Practitioners. Joint Statistical Meeting, 2006.

Gelman, A., Carlin, J. B., Stern, H. S., & Rubin, D. B. (2004). *Bayesian Data Analysis*. Boca Raton, FL: Chapman & Hall/CRC.

Schafer, J. L. (1997). *Analysis of Incomplete Multivariate Data*. Boca Raton, FL: Chapman & Hall/CRC.

*SAS OnlineDoc: Version 8*. Chapter 9, The MI Procedure. http://v8doc.sas.com/sashtml/ Accessed July 7, 2006.